Software VNA and Microwave Network Design and Characterisation

Software VNA and Microwave Network Design and Characterisation

Zhipeng Wu
University of Manchester, UK

John Wiley & Sons, Ltd

To Guoping, William and Richard

Contents

Foreword

Software VNA and Microwave Network Design and Characterisation is a unique contribution to the microwave literature. It fills a need in the education and training of microwave engineers and builds upon well-established texts such as *Fields and Waves in Communication Electronics* by S. Ramo, J. Whinnery and T. Van Duzer, *Foundations for Microwave Engineering* by R.E. Collin and *Microwave Engineering* by D. Pozar. The 'virtual vector network analyser' that can be downloaded from the CD supplied with the book enables those without access to a real instrument to learn how to use a vector network analyser. The many design examples provide opportunities for the reader to become familiar with the Software VNA and the various formats in which frequency responses can be displayed. They also encourage 'virtual experiments'.

Design formulas for many devices are given, but the underlying theory that can be found in other texts is not covered to avoid repetition. A circuit theory or field theory approach is available and this encourages the user to link the two. A novel feature of the book is the introduction and application of a two-port chart that complements the well-known Smith chart, widely used for one-port circuits. The two-port chart enables the frequency response of transmission parameters to be displayed as well as reflection parameters. The range of devices introduced in the book includes stubs, transformers, power dividers/combiners, couplers, filters, antennas and amplifiers. Nonideal behaviour, e.g. the effects of dielectric, conductor and radiation losses, is included for many devices. The devices can be connected to form microwave circuits and the frequency response of the circuit can be 'measured'. The lower frequency limit in the Software VNA is 1 Hz and circuits containing both lumped and distributed devices can be characterised.

Assuming a knowledge of transmission lines, circuits and some electromagnetic theory, Software VNA is suitable for introduction at the final-year undergraduate level and postgraduate levels. Students would be stimulated by the opportunity to 'measure' their own devices and circuits. Experienced microwave engineers will also find Software VNA useful.

L.E. Davis
University of Manchester

Preface

In addition to conventional textbooks, the advances in computer technology and modern microwave test instruments over the past decade have given electrical engineers, researchers and university students two new approaches to study microwave components, devices and circuits. The Vector Network Analyser (VNA) is one of the most desirable instruments in the area of microwave engineering, which can provide fast and accurate characterisation of microwave components, devices or circuits of interest. On the other hand, a commercial microwave circuit simulation software package offers a cost-effective way to study the properties of microwave components and devices before they are used to construct circuits and the properties of the circuits before they are built for testing. However, mainly due to their costs, VNAs and microwave circuit simulators are not widely accessible on a day-to-day basis to many electrical engineers, researchers and university students. This book together with the associated software is intended to fill in the gap between these two aspects with (i) an introduction to microwave network analysis, microwave components and devices, microwave circuit design and (ii) the provision of both device and circuit simulators powered by the analytical formulas published in the literature.

The purpose of the associated software named Software VNA is fourfold. First, it functions as a VNA trainer with a lower frequency limit of 1 Hz and a upper frequency limit of 1000 GHz, providing to those who have not seen or used a VNA before the opportunity to have personal experience of how a VNA would operate in practice and be used for microwave measurements. Secondly, it provides experienced users with an option to get access to the data on a commercial VNA test instrument for data analysis, manipulation or comparison. Thirdly, it provides the users with a simulator equipped with 35 device builders from which an unlimited number of devices can be defined and studied. Analytical CAD equations, many of which have been experimentally verified, are used as models for simulation, giving no hidden numerical errors. The users may also use the Software VNA to verify the limitations and accuracy of the CAD equations. Finally, it provides the users with a circuit simulator that they can use to build circuits and study their properties.

The book has five chapters. In Chapter 1, the basic theory of network analysis is introduced and network parameters are defined. In Chapter 2, the installation and functions of the Software VNA are described. In Chapter 3, the built-in device models are presented with detailed equations and their limitations. In Chapter 4, circuit design and operation principles for impedance matching, impedance transformation, resonators, power dividers, coupler, filters and amplifiers are summarised, and the design examples of these circuits are given in Chapter 5.

The book and its associated software can be used for teaching in the area of microwave engineering.

1

Introduction to Network Analysis of Microwave Circuits

ABSTRACT

Network presentation has been used as a technique in the analysis of low-frequency electrical and electronic circuits. The same technique is equally useful in the analysis of microwave circuits, although different network parameters are used. In this chapter, network parameters for microwave circuit analysis, in particular scattering parameters, are introduced together with a Smith chart for one-port networks and a new chart for two-port networks. The analyses of two-port connected networks and a circuit composed of multi-port networks are also presented.

KEYWORDS

Network analysis, Network parameters, Impedance parameters, Admittance parameters, *ABCD* parameters, Scattering parameters, Smith chart, Two-port chart, Connected networks

Network presentation has been used as a technique in the analysis of low-frequency electrical and electronic circuits (Ramo, Whinnery and van Duzer, 1984). The same technique is equally useful in the analysis of microwave circuits, although different network parameters may be used (Collin, 1966; Dobrowolski, 1991; Dobrowolski and Ostrowski, 1996; Fooks and Zakarev, 1991; Gupta, Garg and Chadha, 1981; Liao, 1990; Montgomery, Dicke and Purcell, 1948; Pozar, 1990; Rizzi, 1988; Ishii 1989; Wolff and Kaul, 1998). Using such a technique, a microwave circuit can be regarded as a network or a composition of a number of networks. Each network may also be composed of many elementary components. A network may have many ports, from which microwave energy flows into or out of the network. One- and two-port networks are, however, the most common, and most commercial network analysers provide measurements for one- or two-port networks. In this chapter, the network analysis will be based on one- and two-port networks. Network parameters, in particular scattering parameters, will be introduced together with a Smith chart for one-port networks and a new chart for two-port networks. The analysis of two connected networks and a circuit composed of a number of networks will also be presented. For further reading, see references at the end of the chapter.

1.1 ONE-PORT NETWORK

A one-port network can be simply represented by load impedance Z to the external circuit. When the network is connected to a sinusoidal voltage source with an open circuit peak voltage V_s and a reference internal impedance of $Z_{0,\mathrm{ref}}$ as shown in Figure 1.1, the circuit can be analysed using the circuit theory based on total voltage and current quantities. It can also be analysed using the transmission-reflection analysis based on incident and reflected voltage and current quantities. Both analyses are described below. The reference internal impedance $Z_{0,\mathrm{ref}}$ of the source is assumed to be 50 Ω throughout this book.

1.1.1 Total Voltage and Current Analyses

Using circuit theory, the voltage V on the load impedance and the current I flowing through it as shown in Figure 1.1 are related by

$$V = ZI \tag{1.1}$$

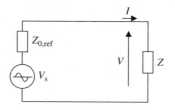

Figure 1.1 Simplified one-port network

and they can be obtained by

$$V = \frac{V_s Z}{Z_{0,\text{ref}} + Z}$$

(1.2a)

and

$$I = \frac{V_s}{Z_{0,\text{ref}} + Z}.$$

(1.2b)

The power delivered to the load impedance by the voltage source can be obtained by

$$P_L = \frac{1}{2}\text{Re}(VI^*) = \frac{|V_s|^2}{2|Z_{0,\text{ref}} + Z|^2}\text{Re}(Z),$$

(1.3)

where * indicates the complex conjugate.

1.1.2   Transmission-Reflection Analysis

1.1.2.1   Voltage and current

Using the transmission-reflection analysis, the incident voltage V^+ is defined to be the voltage that the voltage source could provide to a matched load, i.e. when $Z = Z_{0,\text{ref}}$, and the incident current I^+ to be the corresponding current flowing through the matched load. Hence

$$V^+ = \frac{V_s}{2}$$

(1.4a)

and

$$I^+ = \frac{V^+}{Z_{0,\text{ref}}} = \frac{V_s}{2Z_{0,\text{ref}}}.$$

(1.4b)

Therefore if $Z_L = Z_{0,\text{ref}}$, then $V = V^+$ and $I = I^+$. However, in the general case that $Z \neq Z_{0,\text{ref}}$, the voltage V can be taken to be the superposition of two voltages: the incident voltage V^+ and a reflected voltage V^-. Similarly the current I can be taken as the superposition of two currents: the incident current I^+ and a reflected current I^-. Since V can be written as $V = IZ_{0,\text{ref}} + V_e$ with

$$V_e = \frac{V_s(Z - Z_{0,\text{ref}})}{Z + Z_{0,\text{ref}}},$$

(1.5)

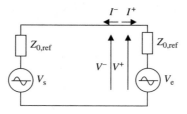

Figure 1.2   Equivalent circuit

the circuit in Figure 1.1 can be represented by an equivalent circuit shown in Figure 1.2. The load impedance Z is replaced by a 'voltage source' V_e with an 'internal impedance' $Z_{0,\text{ref}}$. By using the superposition theorem, the reflected voltage can be taken to be that produced by the equivalent voltage source V_e so that

$$V^- = \frac{V_e}{2} = \frac{V_s}{2} \frac{(Z - Z_{0,\text{ref}})}{(Z + Z_{0,\text{ref}})} \qquad (1.6a)$$

and

$$I^- = \frac{V^-}{Z_{0,\text{ref}}} = \frac{V_s}{2Z_{0,\text{ref}}} \frac{(Z - Z_{0,\text{ref}})}{(Z - Z_{0,\text{ref}})}. \qquad (1.6b)$$

The total voltage and current are then

$$V = V^+ + V^- = V_s \frac{Z_{0,\text{ref}}}{Z + Z_{0,\text{ref}}} \qquad (1.7a)$$

and

$$I = I^+ - I^- = V_s \frac{1}{Z + Z_{0,\text{ref}}}, \qquad (1.7b)$$

which are the same as those in Equation (1.2) obtained using circuit theory.

1.1.2.2   Reflection coefficient

Using the transmission-reflection analysis, the total voltage is expressed as the sum of the incident voltage and the reflected voltage, and the current as the difference of the incident current and the reflected current. For the convenience of analysis, a reflection coefficient can be introduced to relate the reflected and incident quantities. The reflection coefficient is defined as

$$\Gamma = \frac{V^-}{V^+} = \frac{I^-}{I^+} = \frac{Z - Z_{0,\text{ref}}}{Z + Z_{0,\text{ref}}} = \frac{Y_{0,\text{ref}} - Y}{Y_{0,\text{ref}} + Y}, \qquad (1.8)$$

where $Y_{0,\text{ref}} = 1/Z_{0,\text{ref}}$ and $Y = 1/Z$ and is defined with respect to the reference impedance $Z_{0,\text{ref}}$.

The total voltage and current at the load can then be expressed as

$$V = V^+(1 + \Gamma) \tag{1.9a}$$

and

$$I = I^+(1 - \Gamma). \tag{1.9b}$$

Hence the total voltage and current quantities can be obtained when Γ is determined.

1.1.2.3 Power

Associated with the incident voltage V^+ and the incident current I^+ is the incident power which is given by

$$P^+ = \frac{1}{2}\text{Re}(V^+I^{+*}) = \frac{1}{2}\frac{|V^+|^2}{Z_{0,\text{ref}}} = \frac{|V_s|^2}{8Z_{0,\text{ref}}} = P_{\text{max}}. \tag{1.10}$$

This power is also the maximum power available from the voltage source. Similarly the reflected power is associated with the reflected voltage V^- and the reflected current I^- and is given by

$$P^- = \frac{1}{2}\text{Re}(V^-I^{-*}) = \frac{1}{2}\frac{|V^-|^2}{Z_{0,\text{ref}}} = P_{\text{max}}|\Gamma|^2. \tag{1.11}$$

The power delivered to the load impedance is the difference between the incident power and the reflected power, i.e.

$$P_{\text{L}} = P^+ - P^- = P^+(1 - |\Gamma|^2), \tag{1.12}$$

which is identical to Equation (1.3).

1.1.2.4 Introduction of a_1 and b_1

Since the incident power is related to V^+ and $Z_{0,\text{ref}}$ and the reflected power to V^- and $Z_{0,\text{ref}}$ as in Equations (1.10) and (1.11), their expressions can be simplified with the introduction of two new quantities a_1 and b_1 which are defined as (Collin, 1966; Pozar, 1990)

$$a_1 = \frac{V^+}{\sqrt{Z_{0,\text{ref}}}} \tag{1.13a}$$

and

$$b_1 = \frac{V^-}{\sqrt{Z_{0,\text{ref}}}}. \qquad (1.13b)$$

Using these two new quantities, the incident, reflected and total powers can then be expressed, respectively, as

$$P^+ = \frac{1}{2}|a_1|^2, \qquad (1.14a)$$

$$P^- = \frac{1}{2}|b_1|^2 \qquad (1.14b)$$

and

$$P_{\text{L}} = \frac{1}{2}(|a_1|^2 - |b_1|^2). \qquad (1.14c)$$

The voltage and current quantities can also be written as

$$V^+ = a_1\sqrt{Z_{0,\text{ref}}}, \qquad (1.15a)$$

$$V^- = b_1\sqrt{Z_{0,\text{ref}}}, \qquad (1.15b)$$

$$I^+ = \frac{a_1}{\sqrt{Z_{0,\text{ref}}}}, \qquad (1.15c)$$

$$I^- = \frac{b_1}{\sqrt{Z_{0,\text{ref}}}}, \qquad (1.15d)$$

$$V = (a_1 + b_1)\sqrt{Z_{0,\text{ref}}} \qquad (1.15e)$$

and

$$I = \frac{a_1 - b_1}{\sqrt{Z_{0,\text{ref}}}}. \qquad (1.15f)$$

The reflection coefficient defined in Equation (1.8) becomes

$$\Gamma = \frac{b_1}{a_1} = \frac{Z - Z_{0,\text{ref}}}{Z + Z_{0,\text{ref}}} = \frac{Y_{0,\text{ref}} - Y}{Y_{0,\text{ref}} + Y}. \tag{1.16}$$

Using a_1 and b_1, the signal reflection property of the one-port network can be described by

$$b_1 = \Gamma a_1. \tag{1.17}$$

1.1.2.5 Z in terms of Γ

The formulas derived above are useful for the analysis of one-port network when the load impedance is known. However, very often in practice, Z has to be determined from measurement. In this case, Z can be obtained from the measurement of the reflection coefficient using

$$Z = \frac{1 + \Gamma}{1 - \Gamma} Z_{0,\text{ref}}. \tag{1.18}$$

1.1.3 Smith Chart

1.1.3.1 Impedance chart

The impedance Smith chart (Smith, 1939, 1944) is based on the expression of the reflection coefficient Γ in terms of load impedance Z. With the introduction of the normalised load impedance with respect to the reference impedance $Z_{0,\text{ref}}$ as

$$z = \frac{Z}{Z_{0,\text{ref}}} = r + \text{j}x, \tag{1.19}$$

where r and x are the normalised resistance and reactance, respectively, the reflection coefficient Γ can be written as

$$\Gamma = \frac{z - 1}{z + 1} = \frac{r + \text{j}x - 1}{r + \text{j}x + 1} = u + \text{j}v, \tag{1.20}$$

where u and v are the real and imaginary projections of Γ on the complex u–v plane. Equation (1.20) can be rearranged to give the following two separate equations:

$$\left(u - \frac{r}{1 + r} \right)^2 + v^2 = \frac{1}{(1 + r)^2} \tag{1.21a}$$

and

$$(u-1)^2 + \left(v - \frac{1}{x}\right)^2 = \frac{1}{x^2}. \tag{1.21b}$$

Equation (1.21a) represents a family of constant resistance circles. The centre of the circle for a normalised resistance r is at $(r/(1+r), 0)$ and the radius is $1/(1+r)$. Equation (1.21b) describes a family of constant reactance circles. The centre of the circle with a normalised reactance x is at $(1, 1/x)$ and the radius of the circle is $1/|x|$. A simplified impedance Smith chart is shown in Figure 1.3. On the chart, the normalised resistance and reactance values, r and x, can be read when the reflection coefficient Γ is plotted. On the other hand, the complex reflection coefficient can be determined when r and x values are known and plotted on the impedance chart.

1.1.3.2   Admittance chart

With the introduction of the normalised admittance

$$y = \frac{Y}{Y_{0,\text{ref}}} = g + jb, \tag{1.22}$$

where g and b are the normalised conductance and admittance, respectively. Equation (1.16) for the reflection coefficient Γ can be rearranged to

$$-\Gamma = \frac{g + jb - 1}{g + jb + 1}. \tag{1.23}$$

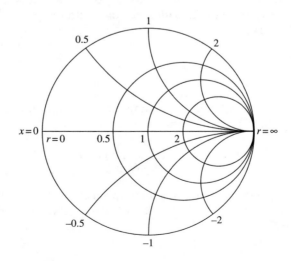

Figure 1.3   Simplified Smith chart

Comparing Equation (1.23) with Equation (1.20), it can be seen that g would be the same as r and x as b if Γ in Equation (1.20) is replaced by $-\Gamma$. Hence the admittance Smith chart can be obtained by rotating the impedance chart by 180°. Alternatively the values of g and b can be read on the impedance chart as for r and x by plotting $-\Gamma_L$ on the impedance chart.

1.1.4 Terminated Transmission Line

For a terminated transmission line, as shown in Figure 1.4, the voltage and current at any position on the transmission line can be described by the following two equations:

$$V(z) = V^+ e^{-\gamma z} + V^- e^{\gamma z} = V^+(e^{-\gamma z} + \Gamma_L e^{\gamma z}) \tag{1.24a}$$

and

$$I(z) = I^+ e^{-\gamma z} - I^- e^{\gamma z} = \frac{V^+}{Z_0}(e^{-\gamma z} - \Gamma_L e^{\gamma z}), \tag{1.24b}$$

where Z_0 is the characteristic impedance of the transmission line, Γ_L the reflection coefficient with respect to the transmission line impedance of Z_0 and γ the propagation constant. The reflection coefficient Γ_L is given by

$$\Gamma_L = \frac{Z_L - Z_0}{Z_L - Z_0} \tag{1.25a}$$

and the propagation constant by

$$\gamma = \alpha + j\beta = (\alpha_c + \alpha_d + \alpha_r) + j\frac{\omega}{v_p}, \tag{1.25b}$$

where α is the attenuation constant in Nepers/m, β the phase constant in rad/m and v_p the phase velocity. The attenuation constant may consist of

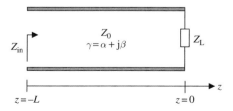

Figure 1.4 Terminated transmission line

three parts: α_c for conducting loss, α_d for dielectric loss and α_r for radiation loss. The dielectric loss α_d can be expressed as

$$\alpha_d = \frac{1}{2}\beta \tan\delta, \tag{1.26}$$

where $\tan\delta$ is the loss tangent of the dielectric material used in the transmission line.

The input impedance of the transmission line of length L can be obtained by

$$Z_{in} = Z_0 \frac{Z_L + Z_0 \tanh(\gamma L)}{Z_0 + Z_L \tanh(\gamma L)}. \tag{1.27}$$

If the terminated transmission line is connected to a sinusoidal voltage source with an internal impedance $Z_{0,ref}$, the reflection coefficient with respect to the voltage source is

$$\Gamma = \frac{Z_{in} - Z_{0,ref}}{Z_{in} - Z_{0,ref}}. \tag{1.28}$$

For a lossless transmission line, the input impedance becomes

$$Z_{in} = Z_0 \frac{Z_L + jZ_0 \tan(\beta L)}{Z_0 + jZ_L \tan(\beta L)}. \tag{1.29}$$

The total voltage and current on the transmission line depend not only on the source, but also on the load impedance. The reflection from the load impedance will cause a standing wave on the transmission line with V_{max} at voltage maximum positions and V_{min} at voltage minimum positions. For a lossless transmission line, the voltage standing wave ratio (VSWR) can be defined as

$$\text{VSWR} = \frac{V_{max}}{V_{min}} = \frac{1 + |\Gamma_L|}{1 - |\Gamma_L|}. \tag{1.30}$$

This equation can be used to determine the amplitude of Γ_L by measuring the VSWR.

1.2   TWO-PORT NETWORK

1.2.1   Total Quantity Network Parameters

The most commonly used total quantity network parameters for two-port networks are Z, Y and $ABCD$ parameters. Using these parameters, the

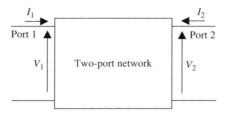

Figure 1.5   A two-port network

properties of the networks are represented by the relation between the total voltage and current quantities at the network ports. For a two-port network, as shown in Figure 1.5, the Z, Y and $ABCD$ parameters are defined, respectively, as (Ramo, Whinnery and van Duzer, 1984)

$$\begin{bmatrix} V_1 \\ V_2 \end{bmatrix} = \begin{bmatrix} Z_{11} & Z_{12} \\ Z_{21} & Z_{22} \end{bmatrix} \begin{bmatrix} I_1 \\ I_2 \end{bmatrix},$$ (1.31a)

$$\begin{bmatrix} I_1 \\ I_2 \end{bmatrix} = \begin{bmatrix} Y_{11} & Y_{12} \\ Y_{21} & Y_{22} \end{bmatrix} \begin{bmatrix} V_1 \\ V_2 \end{bmatrix},$$ (1.31b)

$$\begin{bmatrix} V_1 \\ I_1 \end{bmatrix} = \begin{bmatrix} A & B \\ C & D \end{bmatrix} \begin{bmatrix} V_2 \\ -I_2 \end{bmatrix},$$ (1.31c)

where V_1 and V_2 are the total voltages at Port 1 and Port 2, respectively, and I_1 and I_2 the total currents flowing into the network at Port 1 and Port 2, respectively.

1.2.2   Determination of Z, Y and $ABCD$ Parameters

The network parameters can be determined using open-circuit or short-circuit terminations at the network ports as shown.

The Z parameters can be obtained using open-circuit terminations as

$$Z_{11} = \left. \frac{V_1}{I_1} \right|_{I_2=0} \quad Z_{12} = \left. \frac{V_1}{I_2} \right|_{I_1=0}$$

$$Z_{21} = \left. \frac{V_2}{I_1} \right|_{I_2=0} \quad Z_{22} = \left. \frac{V_2}{I_2} \right|_{I_1=0}$$ (1.32)

Similarly, the Y parameters can be obtained using short-circuit terminations as

$$Y_{11} = \frac{I_1}{V_1}\Big|_{V_2=0} \quad Y_{12} = \frac{I_1}{V_2}\Big|_{V_1=0}$$

$$Y_{21} = \frac{I_2}{V_1}\Big|_{V_2=0} \quad Y_{22} = \frac{I_2}{V_2}\Big|_{V_1=0} \qquad (1.33)$$

The $ABCD$ parameters can be obtained using the combination of open-circuit and short-circuit terminations as

$$A = \frac{V_1}{V_2}\Big|_{I_2=0} \quad B = \frac{V_1}{-I_2}\Big|_{V_2=0}$$

$$C = \frac{I_1}{V_2}\Big|_{I_2=0} \quad D = \frac{I_1}{-I_2}\Big|_{V_2=0} \qquad (1.34)$$

1.2.3   Properties of Z, Y and $ABCD$ Parameters

For a reciprocal network,

$$Z_{12} = Z_{21} \quad Y_{12} = Y_{21} \quad AD - BC = 1. \qquad (1.35)$$

For a symmetrical network,

$$Z_{11} = Z_{22} \quad Y_{11} = Y_{22} \quad A = D. \qquad (1.36)$$

1.2.4   Scattering Parameters

By using Z, Y and $ABCD$ parameters, the properties of a two-port network can be described in terms of total voltages and currents at input and output ports of the network. At low frequencies, the voltages and currents can be easily measured so that the Z, Y and $ABCD$ parameters can be determined. At microwave frequencies, however, the voltage and current are difficult to measure due to lack of instrument and also the fact that voltage and current are not always well defined. However, microwave power is relatively easy to measure. A more satisfactory approach is to use variables relating to the incident and reflected quantities. Scattering parameters are then introduced to describe microwave networks. The scattering parameters are easier to measure than voltage or current.

As for the one-port network discussed in Section 1.1.2, the total voltage at each port can be expressed as the sum of an incident voltage and a 'reflected' voltage, and the total current as the difference of an incident current and a 'reflected' current, i.e.

$$V_1 = V_1^+ + V_1^- \quad I_1 = I_1^+ - I_1^-$$
$$V_2 = V_2^+ + V_2^- \quad I_2 = I_2^+ - I_2^-$$

(1.37)

as shown in Figure 1.6.

In general, the source connected to the network can have different internal impedances. However, throughout this book for all two-port networks, it is considered that the sources connected to the network have the same reference internal impedance of $Z_{0,\mathrm{ref}} = 50\ \Omega$. It should be noted that V_1^- and I_1^- include those produced by V_{s2} and V_2^- and I_2^- include those produced by V_{s1}. The incident and 'reflected' voltages and currents satisfy the relation

$$\frac{V_1^+}{I_1^+} = \frac{V_1^-}{I_1^-} = \frac{V_2^+}{I_2^+} = \frac{V_2^-}{I_2^-} = Z_{0,\mathrm{ref}}.$$

(1.38)

Unlike Z, Y and $ABCD$ parameters, scattering parameters do depend on the reference internal impedance chosen.

With the introduction of a_1, b_1, a_2 and b_2 quantities,

$$a_1 = \frac{V_1^+}{\sqrt{Z_{0,\mathrm{ref}}}} = I_1^+ \sqrt{Z_{0,\mathrm{ref}}}\ b_1 = \frac{V_1^-}{\sqrt{Z_{0,\mathrm{ref}}}} = I_1^- \sqrt{Z_{0,\mathrm{ref}}}$$
$$a_2 = \frac{V_2^+}{\sqrt{Z_{0,\mathrm{ref}}}} = I_2^+ \sqrt{Z_{0,\mathrm{ref}}}\ b_2 = \frac{V_2^-}{\sqrt{Z_{0,\mathrm{ref}}}} = I_2^- \sqrt{Z_{0,\mathrm{ref}}},$$

(1.39)

a new set of parameters, i.e. scattering parameters

$$[S] = \begin{bmatrix} S_{11} & S_{12} \\ S_{21} & S_{22} \end{bmatrix},$$

can be defined. The scattering parameters or S-parameters relate b_1 and b_2 to a_1 and a_2 as (Collin, 1966; Pozar, 1990)

$$\begin{bmatrix} b_1 \\ b_2 \end{bmatrix} = \begin{bmatrix} S_{11} & S_{12} \\ S_{21} & S_{22} \end{bmatrix} \begin{bmatrix} a_1 \\ a_2 \end{bmatrix}.$$

(1.40)

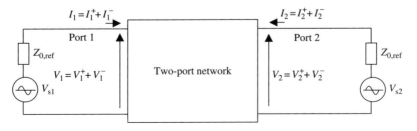

Figure 1.6 A two-port network with external sources

1.2.5   Determination of S-Parameters

The S-parameters can be determined by connecting one of the ports to a source with reference internal impedance $Z_{0,\text{ref}}$ and terminating at the other port with a matched load, i.e. $Z_L = Z_{0,\text{ref}}$, as follows:

$$S_{11} = \frac{b_1}{a_1}\Big|_{a_2=0} \quad S_{12} = \frac{b_1}{a_2}\Big|_{a_1=0}$$

$$S_{21} = \frac{b_2}{a_1}\Big|_{a_2=0} \quad S_{22} = \frac{b_2}{a_2}\Big|_{a_1=0} \tag{1.41}$$

The S-parameters are generally complex parameters and are often represented in terms of amplitude and phase.

1.2.6   Total Voltages and Currents in Terms of a and b Quantities

Using Equation (1.39), the total voltages and currents of the two-port network can be expressed in terms of a_1, a_2, b_1 and b_2 as

$$V_1 = (a_1 + b_1)\sqrt{Z_{0,\text{ref}}} \quad I_1 = (a_1 - b_1)/\sqrt{Z_{0,\text{ref}}}$$
$$V_2 = (a_2 + b_2)\sqrt{Z_{0,\text{ref}}} \quad I_2 = (a_2 - b_2)/\sqrt{Z_{0,\text{ref}}} \tag{1.42}$$

It can be shown that a_1, a_2, b_1 and b_2 can also be written as

$$a_1 = \frac{1}{2}\left(\frac{V_1}{\sqrt{Z_{0,\text{ref}}}} + I_1\sqrt{Z_{0,\text{ref}}}\right) \quad b_1 = \frac{1}{2}\left(\frac{V_1}{\sqrt{Z_{0,\text{ref}}}} - I_1\sqrt{Z_{0,\text{ref}}}\right)$$

$$a_2 = \frac{1}{2}\left(\frac{V_2}{\sqrt{Z_{0,\text{ref}}}} + I_2\sqrt{Z_{0,\text{ref}}}\right) \quad b_2 = \frac{1}{2}\left(\frac{V_2}{\sqrt{Z_{0,\text{ref}}}} - I_2\sqrt{Z_{0,\text{ref}}}\right) \tag{1.43}$$

1.2.7   Power in Terms of a and b Quantities

The incident power at Port 1 and Port 2 is, respectively,

$$P_1^+ = \tfrac{1}{2}|a_1|^2 \quad P_2^+ = \tfrac{1}{2}|a_2|^2 \tag{1.44a}$$

and the 'reflected' power from Port 1 and Port 2 is

$$P_1^- = \tfrac{1}{2}|b_1|^2 \quad P_2^- = \tfrac{1}{2}|b_2|^2. \tag{1.44b}$$

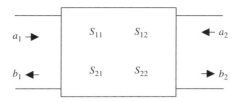

Figure 1.7   A two-port network illustrated using a_1, a_2, b_1 and b_2

The power lost at the network is therefore

$$P_{\text{Loss}} = P_1^+ + P_2^+ - P_1^- P_2^- = \frac{1}{2}(|a_1|^2 + |a_2|^2 - |b_1|^2 - |b_2|^2). \qquad (1.45)$$

Using a_1, a_2, b_1 and b_2 quantities, the power flow of the two-port network can be illustrated as shown in Figure 1.7.

1.2.8   Signal Flow Chart

The power flow can also be represented graphically using a signal flow chart (Pozar, 1990) as shown in Figure 1.8. The chart has four nodes and four branches. The a nodes represent the incoming signals and b nodes the 'reflected' signals. The branches represent the signal flow along the indicated arrow directions, so that Equation (1.40) can be directly written from the graphical representation.

1.2.9   Properties of S-Parameters

The S-parameters have the following properties. $S_{11} = 0$ when Port 1 of the network is matched to the reference internal impedance of the source. Similarly $S_{22} = 0$ when Port 2 of the network is matched to the reference internal impedance of the source.

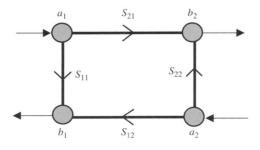

Figure 1.8   Signal flow chart

For a reciprocal two-port network,

$$S_{21} = S_{12}. \tag{1.46a}$$

For a lossless two-port network (Liao, 1990),

$$[S][S^*] = [I] \tag{1.46b}$$

or

$$\begin{aligned}
|S_{11}|^2 + |S_{21}|^2 &= 1 \\
|S_{12}|^2 + |S_{22}|^2 &= 1 \\
S_{11}{}^*S_{12} + S_{21}{}^*S_{22} &= 0 \\
S_{12}{}^*S_{11} + S_{22}{}^*S_{21} &= 0
\end{aligned} \tag{1.46c}$$

where $[I]$ is a unit matrix of rank 2.

1.2.10   Power Flow in a Terminated Two-Port Network

Consider a two-port network connected to a source with a reference internal impedance $Z_{0,\text{ref}}$ as shown in Figure 1.9. The power flow in the terminated two-port network in general depends not only on the S-parameters of the network, but also on the load impedance Z_L. When $Z_L = Z_{0,\text{ref}}$, the power delivered to the load is fully absorbed. No reflection will take place at the termination, i.e. $a_2 = 0$. The returned power from Port 1 of the terminated network is then

$$P_{\text{return}} = P_{\text{max}}|S_{11}|^2, \tag{1.47a}$$

where $P_{\text{max}} = P_1^+$ is the maximum available power from the source, as given in Equation (1.44a), and the power received by the load is

$$P_{\text{load}} = P_{\text{max}}|S_{21}|^2. \tag{1.47b}$$

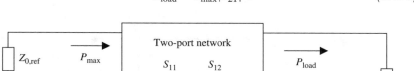
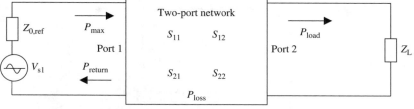

Figure 1.9   A terminated two-port network with an external source connected to Port 1

Hence the input power to the terminated network and the power lost in the network are, respectively,

$$P_{in} = P_{max}(1 - |S_{11}|^2) \qquad (1.48a)$$

and

$$P_{loss} = P_{max}(1 - |S_{11}|^2 - |S_{21}|^2). \qquad (1.48b)$$

The power gain of the two-port network for a matched termination can be obtained as

$$G_m = \frac{P_{load}}{P_{in}} = \frac{|S_{21}|^2}{(1 - |S_{11}|^2)} \qquad (1.49a)$$

and the transducer power gain is given by

$$G_{Tm} = \frac{P_{load}}{P_{max}} = |S_{21}|^2. \qquad (1.49b)$$

The insertion loss due to the two-port network is therefore

$$IL_m = \frac{P_{max}}{P_{load}} = \frac{1}{|S_{21}|^2} \quad \text{or} \quad IL_m(dB) = -20\log_{10}(|S_{21}|). \qquad (1.50)$$

In the general case that $Z_L \neq Z_{0,ref}$, a reflection will take place at the termination, and the reflection coefficient with respect to $Z_{0,ref}$ is

$$\Gamma_L = \frac{Z_L - Z_{0,ref}}{Z_L + Z_{0,ref}}. \qquad (1.51)$$

Using Equation (1.40) and applying Equation (1.17) to the terminated load with $a_2 = \Gamma_L b_2$,

$$b_1 = \left(S_{11} + \frac{\Gamma_L S_{21} S_{12}}{1 - \Gamma_L S_{22}}\right) a_1 \qquad (1.52a)$$

and

$$b_2 = \frac{a_2}{\Gamma_L} = \frac{S_{21}}{1 - \Gamma_L S_{22}} a_1. \qquad (1.52b)$$

can be obtained. Hence the returned power from the terminated network is

$$P_{return} = P_{max} \left| S_{11} + \frac{\Gamma_L S_{21} S_{12}}{1 - \Gamma_L S_{22}} \right|^2 \qquad (1.53a)$$

and the power received by the load, with the consideration of reflection, is

$$P_{\text{load}} = P_{\text{max}} \frac{|S_{21}|^2 (1 - |\Gamma_{\text{L}}|^2)}{|1 - \Gamma_{\text{L}} S_{22}|^2}. \tag{1.53b}$$

The input power to the network is then

$$P_{\text{in}} = P_{\text{max}} \left(1 - \left| S_{11} + \frac{\Gamma_{\text{L}} S_{21} S_{12}}{1 - \Gamma_{\text{L}} S_{22}} \right|^2 \right) \tag{1.54a}$$

and the power lost in the network is

$$P_{\text{loss}} = P_{\text{max}} \left(1 - \left| S_{11} + \frac{\Gamma_{\text{L}} S_{21} S_{12}}{1 - \Gamma_{\text{L}} S_{22}} \right|^2 - \left| \frac{S_{21}}{1 - \Gamma_{\text{L}} S_{22}} \right|^2 \right). \tag{1.54b}$$

With the earlier expression for power on the terminated network, the power gain of the two-port network can be obtained as

$$G = \frac{P_{\text{load}}}{P_{\text{in}}} = \frac{|S_{21}|^2 (1 - |\Gamma_{\text{L}}|^2)}{|1 - S_{22} \Gamma_{\text{L}}|^2 \left(1 - \left| S_{11} + \frac{\Gamma_{\text{L}} S_{21} S_{12}}{1 - S_{22} \Gamma_{\text{L}}} \right|^2 \right)} \tag{1.55a}$$

and the transducer power gain is given by

$$G_{\text{T}} = \frac{P_{\text{load}}}{P_{\text{max}}} = \frac{|S_{21}|^2 (1 - |\Gamma_{\text{L}}|^2)}{|1 - S_{22} \Gamma_{\text{L}}|^2}. \tag{1.55b}$$

The insertion loss due to the two-port network is then

$$\text{IL} = \frac{P_{\text{load,direct}}}{P_{\text{load}}} = \frac{|1 - \Gamma_{\text{L}} S_{22}|^2}{|S_{21}|^2}, \tag{1.56}$$

where $P_{\text{load,direct}}$ is the power received by the load when it is directly connected to the source and $P_{\text{load,direct}} = P_{\text{max}} (1 - |\Gamma_{\text{L}}|)^2$.

1.3   CONVERSIONS BETWEEN Z, Y AND ABCD AND S-PARAMETERS

S-parameters can be converted from Z, Y and ABCD parameters and vice versa. The conversions are shown in Tables 1.1 and 1.2, respectively.

Table 1.1 Conversions from Z, Y and $ABCD$ to S-parameters

Z	Y	$ABCD$
$S_{11} = \dfrac{(Z_{11} - Z_{0,\text{ref}})(Z_{22} + Z_{0,\text{ref}}) - Z_{12}Z_{21}}{(Z_{11} + Z_{0,\text{ref}})(Z_{22} + Z_{0,\text{ref}}) - Z_{12}Z_{21}}$	$S_{11} = \dfrac{(Y_{0,\text{ref}} - Y_{11})(Y_{0,\text{ref}} + Y_{22}) + Y_{12}Y_{21}}{(Y_{0,\text{ref}} + Y_{11})(Y_{0,\text{ref}} + Y_{22}) - Y_{12}Y_{21}}$	$S_{11} = \dfrac{A + B/Z_{0,\text{ref}} - CZ_{0,\text{ref}} - D}{A + B/Z_{0,\text{ref}} + CZ_{0,\text{ref}} + D}$
$S_{12} = \dfrac{2Z_{12}Z_{0,\text{ref}}}{(Z_{11} + Z_{0,\text{ref}})(Z_{22} + Z_{0,\text{ref}}) - Z_{12}Z_{21}}$	$S_{12} = \dfrac{-2Y_{12}Y_{0,\text{ref}}}{(Y_{0,\text{ref}} + Y_{11})(Y_{0,\text{ref}} + Y_{22}) - Y_{12}Y_{21}}$	$S_{12} = \dfrac{2(AD - BC)}{A + B/Z_{0,\text{ref}} + CZ_{0,\text{ref}} + D}$
$S_{21} = \dfrac{2Z_{21}Z_{0,\text{ref}}}{(Z_{11} + Z_{0,\text{ref}})(Z_{22} + Z_{0,\text{ref}}) - Z_{12}Z_{21}}$	$S_{21} = \dfrac{-2Y_{21}Y_{0,\text{ref}}}{(Y_{0,\text{ref}} + Y_{11})(Y_{0,\text{ref}} + Y_{22}) - Y_{12}Y_{21}}$	$S_{12} = \dfrac{2}{A + B/Z_{0,\text{ref}} + CZ_{0,\text{ref}} + D}$
$S_{22} = \dfrac{(Z_{11} + Z_{0,\text{ref}})(Z_{22} - Z_{0,\text{ref}}) - Z_{12}Z_{21}}{(Z_{11} + Z_{0,\text{ref}})(Z_{22} + Z_{0,\text{ref}}) - Z_{12}Z_{21}}$	$S_{22} = \dfrac{(Y_{0,\text{ref}} + Y_{11})(Y_{0,\text{ref}} - Y_{22}) + Y_{12}Y_{21}}{(Y_{0,\text{ref}} + Y_{11})(Y_{0,\text{ref}} + Y_{22}) - Y_{12}Y_{21}}$	$S_{22} = \dfrac{-A + B/Z_{0,\text{ref}} - CZ_{0,\text{ref}} + D}{A + B/Z_{0,\text{ref}} + CZ_{0,\text{ref}} + D}$

Table 1.2 Conversions from S-parameters to Z, Y and ABCD

Z	Y	ABCD
$Z_{11} = Z_{0,\mathrm{ref}} \dfrac{(1+S_{11})(1-S_{22})+S_{12}S_{21}}{(1-S_{11})(1-S_{22})-S_{12}S_{21}}$	$Y_{11} = Y_{0,\mathrm{ref}} \dfrac{(1-S_{11})(1+S_{22})+S_{12}S_{21}}{(1+S_{11})(1+S_{22})-S_{12}S_{21}}$	$A = \dfrac{(1+S_{11})(1-S_{22})+S_{12}S_{21}}{2S_{21}}$
$Z_{12} = Z_{0,\mathrm{ref}} \dfrac{2S_{12}}{(1-S_{11})(1-S_{22})-S_{12}S_{21}}$	$Y_{12} = Y_{0,\mathrm{ref}} \dfrac{-2S_{12}}{(1+S_{11})(1+S_{22})-S_{12}S_{21}}$	$B = Z_{0,\mathrm{ref}} \dfrac{(1+S_{11})(1+S_{22})-S_{12}S_{21}}{2S_{21}}$
$Z_{21} = Z_{0,\mathrm{ref}} \dfrac{2S_{21}}{(1-S_{11})(1-S_{22})-S_{12}S_{21}}$	$Y_{21} = Y_{0,\mathrm{ref}} \dfrac{-2S_{21}}{(1+S_{11})(1+S_{22})-S_{12}S_{21}}$	$C = \dfrac{1}{Z_{0,\mathrm{ref}}} \dfrac{(1-S_{11})(1-S_{22})-S_{12}S_{21}}{2S_{21}}$
$Z_{22} = Z_{0,\mathrm{ref}} \dfrac{(1-S_{11})(1+S_{22})+S_{12}S_{21}}{(1-S_{11})(1-S_{22})-S_{12}S_{21}}$	$Y_{22} = Y_{0,\mathrm{ref}} \dfrac{(1+S_{11})(1-S_{22})+S_{12}S_{21}}{(1+S_{11})(1+S_{22})-S_{12}S_{21}}$	$D = \dfrac{(1-S_{11})(1+S_{22})+S_{12}S_{21}}{2S_{21}}$

1.4   SINGLE IMPEDANCE TWO-PORT NETWORK

A single impedance network in one-port connection has been dealt with in Section 1.1. The consideration here is made for two-port connections. There are two possible two-port configurations as discussed below.

1.4.1 S-Parameters for Single Series Impedance

For a single series impedance two-port network as shown in Figure 1.10, the S-parameters can be obtained as

$$S_{11} = S_{22} = \frac{1}{1 + 2Z_{0,\text{ref}}/Z} = \frac{1}{1 + 2Z_{0,\text{ref}}Y} \qquad (1.57a)$$

and

$$S_{21} = S_{12} = \frac{1}{1 + Z/(2Z_{0,\text{ref}})}, \qquad (1.57b)$$

where Z is the impedance of the series component and $Y = 1/Z$.

1.4.2 S-Parameters for Single Shunt Impedance

For a single shunt impedance two-port network as shown in Figure 1.11, the S-parameters can be written as

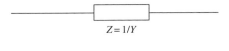

Z = 1/Y

Figure 1.10   Single series impedance two-port network

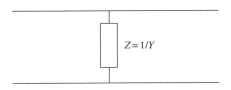

Z = 1/Y

Figure 1.11   Single shunt impedance two-port network

$$-S_{11} = -S_{22} = \frac{1}{1 + 2Y_{0,\text{ref}}Z} = \frac{1}{1 + 2Y_{0,\text{ref}}/Y} \tag{1.58a}$$

and

$$S_{21} = S_{12} = \frac{1}{1 + Y/(2Y_{0,\text{ref}})}, \tag{1.58b}$$

where Z is the impedance of the shunt component and $Y = 1/Z$.

1.4.3 Two-Port Chart

The Smith chart described in Section 1.1.3 is useful for one-port networks where constant resistance and reactance circles can be identified. A similar chart for two-port networks would be useful when a two-port network can be equivalently described by a single impedance network in either series or shunt connections as shown in Figure 1.10 or 1.11. A two-port chart for single impedance network is now introduced (Wu, 2001).

1.4.3.1 Single series impedance network

By introducing S and $(a + jb)$ as defined in Table 1.3, Equations (1.57a) and (1.57b) can be written as

$$S = u + jv = \frac{1}{1 + a + jb} = \frac{1}{A + jB}, \tag{1.59a}$$

where

$$A = 1 + a \text{ and } B = b. \tag{1.59b}$$

Equation (1.59a) can be rearranged to give the following two independent equations:

$$\left(u - \frac{1}{2A}\right)^2 + v^2 = \frac{1}{(2A)^2} \tag{1.60a}$$

Table 1.3 Definition of S and $(a + jb)$ for a single series impedance network

$S = u + jv = |S|e^{j\phi}$	$a + jb$
S_{11} or S_{22}	$(2Z_{0,\text{ref}})Y$
S_{21} or S_{12}	$Z/(2Z_{0,\text{ref}})$

and

$$u^2 + \left(v + \frac{1}{2B}\right)^2 = \frac{1}{(2B)^2}. \qquad (1.60b)$$

Equation (1.60a) represents a family of circles, on each of which $A = 1 + a$ is a constant, with a centre at $(1/(2A), 0)$ and a radius $1/(2A)$. Equation (1.60b) describes a family of circles, on each of which $B = b$ is a constant, with a centre at $(0, 1/(2B))$ and a radius $1/(2B)$. Plotting Equations (1.60a) and (1.60b) gives the two-port chart as shown in Figure 1.12.

For a single series impedance two-port network, the two-port charts for S_{11} and S_{21} are identical. For a passive impedance with $\mathrm{Re}(Z) > 0$, A is greater than or equal to 1. The impedance point on the chart will only be on the right-hand side of the v axis. If S_{11} or S_{21} is plotted on the chart so that the A and B values can be read or determined, the impedance Z or admittance Y can be found from the expressions given in Table 1.4. It can be seen from Table 1.4 that the S_{11} chart is most suitable for finding the admittance Y, but the S_{21} chart is most suitable for obtaining the impedance Z.

1.4.3.2 Single shunt impedance network

With the definition of S and $(a + \mathrm{j}b)$ for a single shunt impedance two-port network, as given in Table 1.5, Equation (1.58a) can be written as

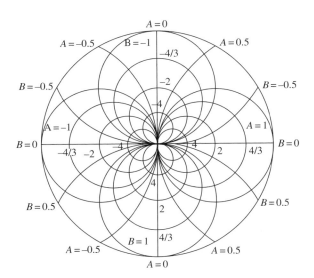

Figure 1.12 The two-port chart for a single impedance network

Table 1.4   Relation between $A + jB$ and Z or Y

$S = |S|e^{j\phi}$	Z or Y
S_{11} or S_{22} chart	$Y = [(A-1) + jB]/(2Z_{0,\text{ref}})$
S_{21} or S_{12} chart	$Z = [(A-1) + jB](2Z_{0,\text{ref}})$

Table 1.5   Definition of S and $(a + jb)$ for a single shunt impedance network

$S = u + jv = |S|e^{j\phi}$	$a + jb$
S_{11} or S_{22}	$(2Y_{0,\text{ref}})Z$
S_{21} or S_{12}	$Y/(2Y_{0,\text{ref}})$

$$-S_{11} = -S_{22} = -(u + jv) = \frac{1}{1 + a + jb} = \frac{1}{A + jB} \qquad (1.61a)$$

and Equation (1.58b) as

$$S_{21} = S_{12} = (u + jv) = \frac{1}{1 + a + jb} = \frac{1}{A + jB}, \qquad (1.61b)$$

where $A = 1 + a$ and $B = b$ as defined in Equation (1.61b). Since Equation (1.61b) is identical to Equation (1.59a), the chart shown in Figure 1.12 is applicable to S_{21} and S_{12}, but with a different definition of $a + jb$ as given in Table 1.5. Equation (1.61.a) is, however, different from Equation (1.59a). It can be rearranged to give the following two independent equations:

$$\left(u + \frac{1}{2A}\right)^2 + v^2 = \frac{1}{(2A)^2} \qquad (1.62a)$$

and

$$u^2 + \left(v - \frac{1}{2B}\right)^2 = \frac{1}{(2B)^2}. \qquad (1.62b)$$

Equations (1.62a) and (1.62b), respectively, represent the same families of circles as Equations (1.60a) and (1.60b) except that the constant A circles now have a centre at $(-1/(2A), 0)$ and constant B circles a centre at $(0, -1/(2B))$ and a radius $1/2B$. Hence the chart shown in Figure 1.12 is generally valid for a single shunt impedance two-port network, but with the understanding that A and B change signs for the S_{11} or S_{22} chart, and $(a + jb)$ is defined in Table 1.5 rather than in Table 1.3.

For a single shunt impedance two-port network, the two-port charts for S_{11} and S_{21} have 180° rotational symmetry. This is different from that for a single series impedance two-port network where the S_{11} and S_{21} charts are identical. This property can be used to identify whether the single impedance in two-port network is in series or shunt connection. For a passive impedance with $\mathrm{Re}(Z) > 0$, A is greater than or equal to 1. The impedance point on the S_{11} chart will only be on the left-hand side of the v axis. If S_{11} or S_{21} is plotted on the chart so that the A and B values can be read or determined, the impedance Z or admittance Y can be found from the expression in Table 1.6. It can be seen from Table 1.6 that the S_{11} chart is most suitable for finding the impedance Z, but the S_{21} chart is most suitable for obtaining the admittance Y.

1.4.3.3   Scaling property

In the two-port chart shown in Figure 1.12, which can be used as either transmission or reflection chart, the outermost circle corresponds to $|S| = 1$, i.e. the chart has a unity scale. Hence for small values of S-parameters, the points plotted on the chart will be concentrated at the centre of the chart. On the other hand, the amplitude of the S-parameter may be greater than unity. It is therefore useful to be able to change the scale of the chart so that the S-parameter of interest can be better displayed. This can be done using the scaling property of the two-port chart as described below.

If the scale of the outermost circle is increased or decreased from unity to M so that on the circle $|S| = M$, the length of the displayed S-parameter, $S = u + \mathrm{j}v$, on the chart will decrease or increase accordingly. The S-parameter is thus scaled to $S' = (1/M) = U + \mathrm{j}V$ with $U = (1/M)u$ and $V = (1/M)v$. It can be shown using Equations (1.59a), (1.61a) and (1.61b) that U and V satisfy the following equations:

$$\left(U \pm \frac{1}{2A'}\right)^2 + V^2 = \frac{1}{(2A')^2} \tag{1.63a}$$

and

$$U^2 + \left(V \pm \frac{1}{2B'}\right)^2 = \frac{1}{(2B')^2}, \tag{1.63b}$$

Table 1.6   Relation between $A + \mathrm{j}B$ and Z or Y

$S = |S|e^{\mathrm{j}\phi}$	Z or Y
S_{11} or S_{22} chart	$Z = [(A-1) + \mathrm{j}B]/(2Y_{0,\mathrm{ref}})$
S_{21} or S_{12} chart	$Y = [(A-1) + \mathrm{j}B](2Y_{0,\mathrm{ref}})$

where

$$A' = AM, \ B' = BM \qquad (1.64a)$$

and

$$A = \frac{A'}{M}, \ B = \frac{B'}{M}. \qquad (1.64b)$$

Equation (1.63a) is identical to Equation (1.60a) or (1.62a) and Equation (1.63b) to Equation (1.60b) or (1.62b). Hence, the same chart can be used for the scaled S-parameter, with the scale of the chart changed to the amplitude of S-parameter represented at the outermost circle. The A' and B' values for the scaled S-parameter can be read on the unity chart as if there were no scaling. The actual values of A and B can be obtained using Equation (1.64b). Alternatively when the scale is changed from unity, the A and B values shown on the chart are updated to the corresponding A' and B' values as given in Equation (1.64a). The values read on the chart will then be the A and B values directly. The impedance or admittance values, i.e. Z or Y, can be found using the expressions in Table 1.4 or 1.6.

1.4.4   Applications of Two-Port Chart

1.4.4.1   Identification of pure resonance

For a pure RLC resonance, $(a+jb)$ is related to the resonance frequency f_0 and the unloaded quality factor Q_0. Assuming the following equation can be established from Table 1.3 or 1.5 for a given resonance,

$$a+jb = a\left(1+jQ_0\left(\frac{f}{f_0}-\frac{f_0}{f}\right)\right) = a(1+jQ_0\delta(f)), \qquad (1.65)$$

Equations (1.59a), (1.61a) and (1.61b) can be written as

$$\pm S(f) = \pm(u+jv) = \frac{1}{A+jB} = \frac{1/A}{1+jB/A} = \frac{S_0}{1+jQ_L\delta(f)}, \qquad (1.66a)$$

where the '$-$' sign applies to Equation (1.59a),

$$S_0 = \pm S(f_0) = \frac{1}{1+a} = \frac{1}{A} \qquad (1.66b)$$

and Q_L is the loaded quality factor and

$$Q_L = (1 - |S_0|)Q_0 \text{ or } Q_0 = \frac{Q_L}{(1 - |S_0|)}. \qquad (1.66c)$$

For a pure resonance, A is a constant and B changes with frequency. Hence the trace of the resonance on a two-port chart discussed in Section 1.4.3 would be a pure circle on which $A = 1 + a$ as the frequency changes from 0 to ∞. This property can be used to identify whether the resonance is a pure RLC resonance or not.

1.4.4.2 Q-factor measurements

For a pure resonance, as indicated in Equation (1.66a), the loaded quality factor Q_L can be measured at frequencies f_1 and f_2 on the $A = \pm B$ lines. At these frequencies,

$$Q_L \delta(f_{1,2}) = 1, \qquad (1.67)$$

i.e. $|S(f_1)|^2$ and $|S(f_2)|^2$ fall to one-half of the $|S_0|^2$ value at the resonance frequency f_0, and $S(f_1)$ and $S(f_2)$ have a phase shift of 45° from the phase of $S(f_0)$ at the resonance frequency. With a further measurement of $|S_0|$, the unloaded quality factor Q_0 can be obtained using Equation (1.66c). The only parameter that remains to be defined is the suitable S-parameter. The S-parameter valid for the assumption in Equation (1.65) leading to Equation (1.66) is listed in Table 1.7.

1.4.4.3 Resonance with power-dependent losses

When the resistance or its equivalence in the RLC series/parallel circuit is power dependent or nonlinear with respect to the power loss on the resonance circuit which happens, for example, in a high-temperature

Table 1.7 Suitable S-parameter for Q-factor measurement

	Single series impedance two-port network		Single shunt impedance two-port network	
Resonance type	RLC series	RLC parallel	RLC series	RLC parallel
S_{21} response	Bandpass	Bandstop	Bandstop	Bandpass
S-parameter for Q-factor measurement	S_{21} or S_{12}	S_{11} or S_{22}	S_{11} or S_{22}	S_{21} or S_{12}

superconducting resonator, the power-dependent resonance can be observed from the two-port chart. The trace of the resonance will be symmetrical about the v axis at frequencies around f_0, but will divert from the constant A circle. Such a property can be used to identify power-dependent resonance of a resonator or a resonant circuit.

1.4.4.4   Impedance or admittance measurement using the two-port chart

The S_{21} chart may be used as an alternative to the Smith chart for the load impedance or admittance measurement of a one-port network. In this case, the transmitted signal is measured, rather than the reflected signal, which may give practical advantages.

The series configuration shown in Figure 1.10 can be used for the impedance measurement using the S_{21} chart. The impedance $Z = R_Z + jX_Z$ can be obtained from the chart using

$$R_Z = (A - 1)(2Z_{0,\mathrm{ref}}) \text{ and } X_Z = 2Z_{0,\mathrm{ref}}B. \tag{1.68}$$

The shunt configuration shown in Figure 1.11 can be used for the admittance measurement using the S_{21} chart. The admittance $Y = G_Y + jB_Y$ can be obtained from the chart using

$$G_Y = (A - 1)(2Y_{0,\mathrm{ref}}) \text{ and } B_Y = B(2Y_{0,\mathrm{ref}}). \tag{1.69}$$

1.5   S-PARAMETERS OF COMMON ONE- AND TWO-PORT NETWORKS

For convenience, the S-parameter for a number of common one- and two-port networks are given in Table 1.8.

1.6   CONNECTED TWO-PORT NETWORKS

1.6.1   T-Junction

When one arm of the T-junction is connected to a shunt impedance or a network with a reflection coefficient Γ_T with respect to $Z_{0,\mathrm{ref}}$ forming a two-port network as shown in Figure 1.13, the resultant S-parameters are given by

$$[S] = \frac{\begin{bmatrix} \Gamma_T - 1 & 2(1 + \Gamma_T) \\ 2(1 + \Gamma_T) & \Gamma_T - 1 \end{bmatrix}}{3 + \Gamma_T}. \tag{1.70}$$

Table 1.8   *S*-parameters of common one- and two-port networks

Network	S-parameters
$Z = 1/Y$	$\Gamma_L = \dfrac{Z_L - Z_{0,\text{ref}}}{Z_L + Z_{0,\text{ref}}} = \dfrac{Y_{0,\text{ref}} - Y_L}{Y_{0,\text{ref}} + Y_L}$
Z	$[S] = \dfrac{\begin{bmatrix} Z & 2Z_{0,\text{ref}} \\ 2Z_{0,\text{ref}} & Z \end{bmatrix}}{Z + 2Z_{0,\text{ref}}}$
Y	$[S] = \dfrac{\begin{bmatrix} -Y & 2Y_{0,\text{ref}} \\ 2Y_{0,\text{ref}} & -Y \end{bmatrix}}{Y + 2Y_{0,\text{ref}}}$
$Z_1 \quad Z_2$ Z_3	$[S] = \dfrac{\begin{bmatrix} -Z_{0,\text{ref}}^2 + (Z_1 - Z_2)Z_{0,\text{ref}} + Z_1 Z_2 + Z_2 Z_3 + Z_3 Z_1 & 2Z_{0,\text{ref}}Z_3 \\ 2Z_{0,\text{ref}}Z_3 & -Z_{0,\text{ref}}^2 + (Z_2 - Z_1)Z_{0,\text{ref}} + Z_1 Z_2 + Z_2 Z_3 + Z_3 Z_1 \end{bmatrix}}{Z_{0,\text{ref}}^2 + (Z_1 + Z_2 + 2Z_3)Z_{0,\text{ref}} + Z_1 Z_2 + Z_2 Z_3 + Z_3 Z_1}$
Y_2 $Y_1 \quad Y_3$	$[S] = \dfrac{\begin{bmatrix} -Y_{0,\text{ref}}^2 - (Y_1 - Y_2)Y_{0,\text{ref}} - (Y_1 Y_2 + Y_2 Y_3 + Y_3 Y_1) & 2Y_{0,\text{ref}}Y_3 \\ 2Y_{0,\text{ref}}Y_3 & Y_{0,\text{ref}}^2 - (Y_2 - Y_1)Y_{0,\text{ref}} - (Y_1 Y_2 + Y_2 Y_3 + Y_3 Y) \end{bmatrix}}{Y_{0,\text{ref}}^2 + (Y_1 + Y_2 + 2Y_3)Y_{0,\text{ref}} + Y_1 Y_2 + Y_2 Y_3 + Y_3 Y_1}$
$1:n$	$[S] = \dfrac{\begin{bmatrix} 1 - n^2 & 2n \\ 2n & n^2 - 1 \end{bmatrix}}{1 + n^2}$
$n:1$	$[S] = \dfrac{\begin{bmatrix} n^2 - 1 & 2n \\ 2n & 1 - n^2 \end{bmatrix}}{1 + n^2}$
Gain block or attenuator $\pm A_{\text{dB}}$	$[S] = \begin{bmatrix} 10^{\pm A_{\text{dB}}/20} & 0 \\ 0 & 10^{\pm A_{\text{dB}}/20} \end{bmatrix}$
Isolator	$[S] = \begin{bmatrix} 0 & 0 \\ 1 & 0 \end{bmatrix}$

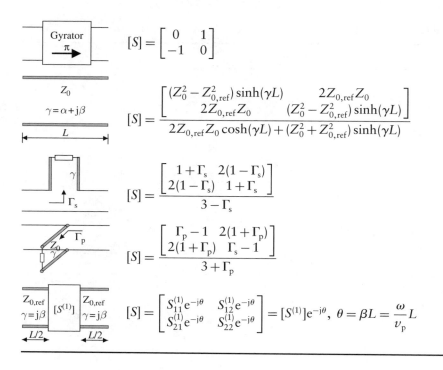

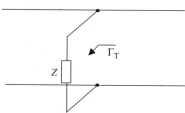

Figure 1.13   T-junction connection

1.6.2   Cascaded Two-Port Networks

For two two-port networks in cascade as shown in Figure 1.14, the resultant S-parameters are

$$[S] = \begin{bmatrix} S_{11}^{(1)} + \dfrac{S_{12}^{(1)} S_{21}^{(1)} S_{11}^{(2)}}{1 - S_{22}^{(1)} S_{11}^{(2)}} & \dfrac{S_{12}^{(1)} S_{12}^{(2)}}{1 - S_{22}^{(1)} S_{11}^{(2)}} \\[3ex] \dfrac{S_{21}^{(1)} S_{21}^{(2)}}{1 - S_{22}^{(1)} S_{11}^{(2)}} & S_{22}^{(2)} + \dfrac{S_{12}^{(2)} S_{21}^{(2)} S_{11}^{(1)}}{1 - S_{22}^{(1)} S_{11}^{(2)}} \end{bmatrix}. \tag{1.71}$$

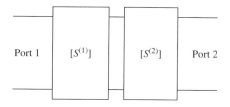

Figure 1.14 Two two-port networks in cascade

When more than two two-port networks are connected in cascade, Equation (1.71) can be used repeatedly to obtain the resultant S-parameters of the cascaded networks.

1.6.3   Two-Port Networks in Series and Parallel Connections

In addition to the cascade connection, two-port networks can also be connected in series or parallel or their combinations, forming a new two-port network. Four configurations are shown in Figure 1.15.

It has been shown that the resultant S-parameters for the above four configurations can be expressed as (Bodharamik, Besser and Newcomb, 1971)

$$[S] = [E]h([E][S^{(1)}], [E][S^{(2)}]), \tag{1.72a}$$

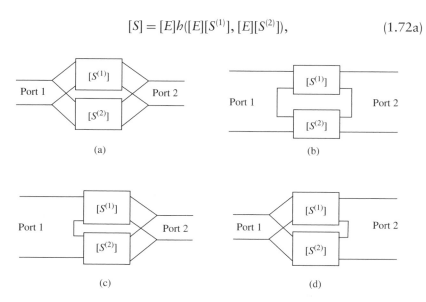

Figure 1.15 Two-port network connection in different configurations: (a) parallel to parallel, (b) series to series, (c) series to parallel and (d) parallel to series

Table 1.9   Matrix [E] for different connections

Port 1	Port 2	[E]
Parallel	Parallel	$\begin{bmatrix} 1 & 0 \\ 0 & 1 \end{bmatrix}$
Series	Series	$\begin{bmatrix} -1 & 0 \\ 0 & -1 \end{bmatrix}$
Series	Parallel	$\begin{bmatrix} -1 & 0 \\ 0 & 1 \end{bmatrix}$
Parallel	Series	$\begin{bmatrix} 1 & 0 \\ 0 & -1 \end{bmatrix}$

where

$$h([S_1], [S_2]) = [A]^{-1}\{[B] + 4[C][S_2]([A] - [B][S_2])^{-1}[C]\} \tag{1.72b}$$

with

$$[A] = 3[I] + [S_1], \tag{1.72c}$$

$$[B] = [S_1] - [I], \tag{1.72d}$$

$$[C] = [S_1] + [I], \tag{1.72e}$$

$$[I] = \begin{bmatrix} 1 & 0 \\ 0 & 1 \end{bmatrix} \tag{1.72f}$$

and [E] given in Table 1.9.

1.7   SCATTERING MATRIX OF MICROWAVE CIRCUITS COMPOSED OF ONE-PORT AND MULTI-PORT DEVICES

1.7.1   S-Parameters of a Multi-Port Device

The S-parameters defined in Section 1.2.4 is for two-port networks. The analysis can be extended to three or more multi-port devices. The total voltage at the nth port can be expressed as the sum of an incident voltage and a 'reflected' voltage and the total current as the difference of an incident current and a 'reflected' current, i.e.

$$V_n = V_n^+ + V_n^- \quad \text{and} \quad I_n = I_n^+ - I_n^-. \tag{1.73}$$

Consider that the sources connected to the multi-port network have the same reference internal impedance of $Z_{0,\text{ref}} = 50\,\Omega$. The incident and 'reflected' voltages and currents satisfy the relation

$$\frac{V_n^+}{I_n^+} = \frac{V_n^-}{I_n^-} = Z_{0,\text{ref}}. \tag{1.74}$$

In the same way as a_1, b_1, a_2 and b_2 are introduced for two-port networks, a_n and b_n can be introduced to the nth port as

$$a_n = \frac{V_n^+}{\sqrt{Z_{0,\text{ref}}}} = I_n^+ \sqrt{Z_{0,\text{ref}}} \quad \text{and} \quad b_n = \frac{V_n^-}{\sqrt{Z_{0,\text{ref}}}} = I_n^- \sqrt{Z_{0,\text{ref}}} \tag{1.75}$$

so that a new set of scattering parameters can be defined, i.e.

$$[S] = \begin{bmatrix} S_{11} & S_{12} & S_{1n} & S_{1N} \\ S_{21} & S_{22} & S_{2n} & S_{2N} \\ S_{n1} & S_{n2} & S_{nn} & S_{nN} \\ S_{N1} & S_{N2} & S_{Nn} & S_{NN} \end{bmatrix} \tag{1.76}$$

for $n = 1$ to N. The multi-port S-parameters relate b_n to a_n by (Collin, 1966; Pozar, 1990)

$$\begin{bmatrix} b_1 \\ b_2 \\ b_n \\ b_N \end{bmatrix} = \begin{bmatrix} S_{11} & S_{12} & S_{1n} & S_{1N} \\ S_{21} & S_{22} & S_{2n} & S_{2N} \\ S_{n1} & S_{n1} & S_{nn} & S_{nN} \\ S_{N1} & S_{N1} & S_{N1} & S_{NN} \end{bmatrix} \begin{bmatrix} a_1 \\ a_2 \\ a_n \\ a_N \end{bmatrix}. \tag{1.77}$$

1.7.2   S-Parameters of a Microwave Circuit

In general, a microwave circuit can be composed of a number of one-port, two-port or multi-port devices, as shown in Figure 1.16. Among the ports, one is connected to an external source with internal impedance $Z_s = Z_{0,\text{ref}}$, one may be connected to an external load impedance $Z_L = Z_{0,\text{ref}}$ and the rest are internally connected. Hence the ports that are connected to external source or load impedance are referred to as the external ports and the rest as the internal ports.

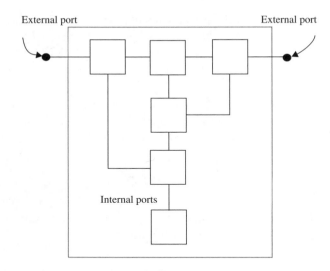

Figure 1.16   Microwave circuit composed of multi-port devices

The signal flow of the circuit can be expressed in terms of S-parameters as $b = Sa$, which can be arranged with the grouping of external and internal ports to

$$\begin{bmatrix} b_e \\ b_i \end{bmatrix} = \begin{bmatrix} S_{ee} & S_{ei} \\ S_{ie} & S_{ii} \end{bmatrix} \begin{bmatrix} a_e \\ a_i \end{bmatrix}, \tag{1.78}$$

where e represents the external ports and i the internal ports. The connections between the ports can be expressed as $b = \Gamma_c a$, where Γ_c is the connection matrix with $\Gamma_{mm} = 0$. The external and internal ports can also be regrouped so that for internally connected ports,

$$b_i = \Gamma_{ii} a_i, \tag{1.79}$$

where $\Gamma_{mn} = 1$, $\Gamma_{nm} = 1$, $a_m = b_n$ and $b_m = a_n$ if ports m and n are connected. Hence

$$a_i = (\Gamma_{ii} - S_{ii})^{-1} S_{ie} a_e \tag{1.80}$$

and

$$b_e = [S_{ee} + S_{ei} (\Gamma_{ii} - S_{ii})^{-1} S_{ie}] a_e \tag{1.81}$$

can be obtained. The new scattering matrix with respect to the external ports is (Dobrowolski, 1991)

$$S_e = S_{ee} + S_{ei} (\Gamma_{ii} - S_{ii})^{-1} S_{ie}. \tag{1.82}$$

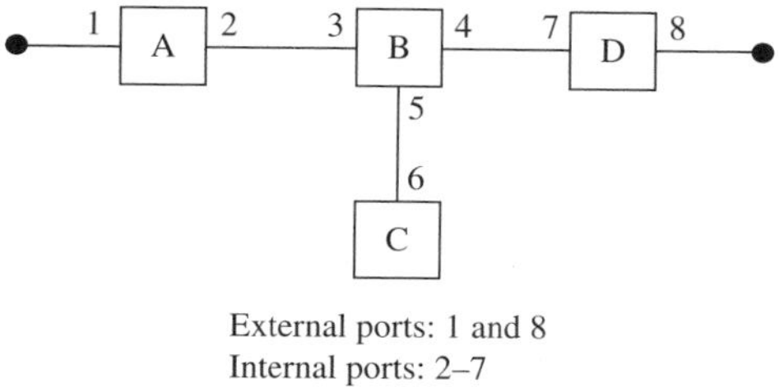

External ports: 1 and 8
Internal ports: 2–7

Figure 1.17   Example of a microwave circuit

For example, the circuit shown in Figure 1.17 consists of four circuit blocks or networks A, B, C and D. Among the numbered ports, 1 and 8 are the external ports and 2–7 are the internal ports. The networks in the circuit have the following S-parameters:

$$[S_A] = \begin{bmatrix} S_{11} & S_{12} \\ S_{21} & S_{22} \end{bmatrix}; \ [S_B] = \begin{bmatrix} S_{33} & S_{34} & S_{35} \\ S_{43} & S_{44} & S_{45} \\ S_{53} & S_{54} & S_{55} \end{bmatrix}; \ [S_C] = [S_{66}]; \ [S_D] = \begin{bmatrix} S_{77} & S_{78} \\ S_{87} & S_{88} \end{bmatrix}.$$

(1.83)

The signal flow of the circuit can be described by

$$\begin{bmatrix} b_1 \\ b_2 \\ b_3 \\ b_4 \\ b_5 \\ b_6 \\ b_7 \\ b_8 \end{bmatrix} = \begin{bmatrix} S_{11} & S_{12} & 0 & 0 & 0 & 0 & 0 & 0 \\ S_{21} & S_{22} & 0 & 0 & 0 & 0 & 0 & S_{87} \\ 0 & 0 & S_{33} & S_{34} & S_{35} & 0 & 0 & 0 \\ 0 & 0 & S_{43} & S_{44} & S_{45} & 0 & 0 & 0 \\ 0 & 0 & S_{53} & S_{54} & S_{55} & 0 & 0 & 0 \\ 0 & 0 & 0 & 0 & 0 & S_{66} & 0 & 0 \\ 0 & 0 & 0 & 0 & 0 & 0 & S_{77} & S_{78} \\ 0 & S_{78} & 0 & 0 & 0 & 0 & S_{87} & S_{88} \end{bmatrix} \begin{bmatrix} a_1 \\ a_2 \\ a_3 \\ a_4 \\ a_5 \\ a_6 \\ a_7 \\ a_8 \end{bmatrix}$$

(1.84)

and the connection matrix of the circuit is

$$\Gamma_c = \begin{matrix} & \begin{matrix} 1 & 2 & 3 & 4 & 5 & 6 & 7 & 8 \end{matrix} \\ \begin{matrix} 1 \\ 2 \\ 3 \\ 4 \\ 5 \\ 6 \\ 7 \\ 8 \end{matrix} & \begin{bmatrix} 0 & 0 & 0 & 0 & 0 & 0 & 0 & 0 \\ 0 & 0 & 1 & 0 & 0 & 0 & 0 & 0 \\ 0 & 1 & 0 & 0 & 0 & 0 & 0 & 0 \\ 0 & 0 & 0 & 0 & 0 & 0 & 1 & 0 \\ 0 & 0 & 0 & 0 & 0 & 1 & 0 & 0 \\ 0 & 0 & 0 & 0 & 1 & 0 & 0 & 0 \\ 0 & 0 & 0 & 1 & 0 & 0 & 0 & 0 \\ 0 & 0 & 0 & 0 & 0 & 0 & 0 & 0 \end{bmatrix} \end{matrix}$$

(1.85)

Rearranging the external and internal ports gives

$$
\begin{bmatrix} b_1 \\ b_8 \\ b_2 \\ b_3 \\ b_4 \\ b_5 \\ b_6 \\ b_7 \end{bmatrix} = \begin{bmatrix} S_{11} & 0 & S_{12} & 0 & 0 & 0 & 0 & 0 \\ 0 & S_{88} & 0 & 0 & 0 & 0 & 0 & S_{87} \\ S_{21} & 0 & S_{22} & 0 & 0 & 0 & 0 & 0 \\ 0 & 0 & 0 & S_{33} & S_{34} & S_{35} & 0 & 0 \\ 0 & 0 & 0 & S_{43} & S_{44} & S_{45} & 0 & 0 \\ 0 & 0 & 0 & S_{53} & S_{54} & S_{55} & 0 & 0 \\ 0 & 0 & 0 & 0 & 0 & 0 & S_{66} & 0 \\ 0 & S_{78} & 0 & 0 & 0 & 0 & 0 & S_{77} \end{bmatrix} \begin{bmatrix} a_1 \\ a_8 \\ a_2 \\ a_3 \\ a_4 \\ a_5 \\ a_6 \\ a_7 \end{bmatrix} \tag{1.86}
$$

and

$$
\Gamma_c = \begin{matrix} & \begin{matrix} 1 & 8 & 2 & 3 & 4 & 5 & 6 & 7 \end{matrix} \\ \begin{matrix} 1 \\ 8 \\ 2 \\ 3 \\ 4 \\ 5 \\ 6 \\ 7 \end{matrix} & \begin{bmatrix} 0 & 0 & 0 & 0 & 0 & 0 & 0 & 0 \\ 0 & 0 & 0 & 0 & 0 & 0 & 0 & 0 \\ 0 & 0 & 0 & 1 & 0 & 0 & 0 & 0 \\ 0 & 0 & 1 & 0 & 0 & 0 & 0 & 0 \\ 0 & 0 & 0 & 0 & 0 & 0 & 0 & 1 \\ 0 & 0 & 0 & 0 & 0 & 0 & 1 & 0 \\ 0 & 0 & 0 & 0 & 0 & 1 & 0 & 0 \\ 0 & 0 & 0 & 0 & 1 & 0 & 0 & 0 \end{bmatrix} \end{matrix} = \begin{bmatrix} \Gamma_{ee} & \Gamma_{ei} \\ \Gamma_{ie} & \Gamma_{ii} \end{bmatrix} \tag{1.87}
$$

so that

$$
S_{ee} = \begin{bmatrix} S_{11} & 0 \\ 0 & S_{88} \end{bmatrix}; \quad S_{ei} = \begin{bmatrix} S_{12} & 0 & 0 & 0 & 0 & 0 \\ 0 & 0 & 0 & 0 & 0 & S_{87} \end{bmatrix}; \quad S_{ie} = \begin{bmatrix} S_{21} & 0 \\ 0 & 0 \\ 0 & 0 \\ 0 & 0 \\ 0 & 0 \\ 0 & S_{78} \end{bmatrix};
$$

$$
S_{ii} = \begin{bmatrix} S_{22} & 0 & 0 & 0 & 0 & 0 \\ 0 & S_{33} & S_{34} & S_{35} & 0 & 0 \\ 0 & S_{43} & S_{44} & S_{45} & 0 & 0 \\ 0 & S_{53} & S_{54} & S_{55} & 0 & 0 \\ 0 & 0 & 0 & 0 & S_{66} & 0 \\ 0 & 0 & 0 & 0 & 0 & S_{77} \end{bmatrix}
$$

$$\tag{1.88}$$

and

$$\Gamma_{ii} = \begin{bmatrix} 0 & 1 & 0 & 0 & 0 & 0 \\ 1 & 0 & 0 & 0 & 0 & 0 \\ 0 & 0 & 0 & 0 & 0 & 1 \\ 0 & 0 & 0 & 0 & 1 & 0 \\ 0 & 0 & 0 & 1 & 0 & 0 \\ 0 & 0 & 1 & 0 & 0 & 0 \end{bmatrix}. \tag{1.89}$$

With the above matrixes, the S-parameters of the circuit, which is a two-port network, can be obtained using

$$S_e = S_{ee} + S_{ei}\,(\Gamma_{ii} - S_{ii})^{-1}\,S_{ie}. \tag{1.90}$$

REFERENCES

Bodharamik, P., Besser, L. and Newcomb, R.W. (1971) 'Two scattering matrix programs for active circuit analysis', *IEEE Transactions on Circuit Theory*, C-18 (6), 610–9.

Collin, R.E. (1966) *Foundations for Microwave Engineering*, McGraw-Hill, New York.

Dobrowolski, J.A. (1991) *Introduction to Computer Methods for Microwave Circuit Analysis and Design*, Artech House, Boston.

Dobrowolski, J.A. and Ostrowski, W. (1996) *Computer Aided Analysis, Modelling, and Design of Microwave Networks: The Wave Approach, Boston*, Artech House, Boston.

Fooks, H.E. and Zakarev, R.A. (1991) *Microwave Engineering Using Circuits*, Prentice Hall, London.

Gupta, K.C., Garg, R. and Chadha, R. (1981) *Computer Aided Design of Microwave Circuits*, Artech House, Boston.

Ishii, T.K. (1989) *Microwave Engineering*, Harcourt Brace Jovanovich, London.

Liao, S.Y. (1990) *Microwave Devices and Circuits*, 3rd edn, Prentice Hall, London.

Montgomery, C.G., Dicke, R.H. and Purcell, E.M. (1948) *Principles of Microwave Circuits*, Vol. 8 of MITRad. Lab. Series, McGraw-Hill, New York.

Pozar, D.M. (1990) *Microwave Engineering*, Addison-Wesley, New York.

Ramo, S., Whinnery, T.R. and van Duzer, T. (1984) *Fields and Waves in Communication Electronics*, 2nd edn, John Wiley & Sons, Ltd, New York.

Rizzi, P.A. (1988) *Microwave Engineering: Passive Circuits*, Prentice Hall, Englewood Cliffs, NJ.

Smith, P.H. (1939) 'Transmission line calculator', *Electronics*, **12** (1), 29–31.

Smith, P.H. (1944) 'An improved transmission line calculator', *Electronics*, **17** (1), 130–3. See also 318–25.

Wolff, E.A. and Kaul, R. (1988) *Microwave Engineering and Systems Applications*, John Wiley & Sons, Ltd, New York.

Wu, Z. (2001) Transmission and reflection charts for two-port single impedance networks. *IEE Proceedings: Microwaves, Antennas and Propagation*, **146** (6), 351–6.

2

Introduction to Software VNA

ABSTRACT

In addition to practical measurements, the *S*-parameters of microwave devices and circuits can be studied using Software VNA. The Software VNA not only simulates many functions of actual commercially available vector network analysers, but also uses the power of simulation to provide additional features that are not available on the actual instruments. In this chapter, the installation procedures of the software package are described. The menu, functions and built-in device categories of the Software VNA are introduced. The use of the integrated Circuit Simulator in the software package for circuit simulation is also illustrated.

KEYWORDS

Software VNA, Software installation, Key functions, Device under test, Circuit simulator, Simulation example

The *S*-parameters described in Chapter 1 can be measured in practice for individual devices or components using a commercial vector network analyser. However, it is not always cost effective to use an experimental approach to carry out a fundamental study of microwave components and devices. Such a study can be carried out using the Software VNA described in this chapter. The Software VNA not only simulates many functions of actual commercially available vector network analysers, but

Software VNA and Microwave Network Design and Characterisation   Zhipeng Wu
© 2007 John Wiley & Sons, Ltd

also uses the power of simulation to provide additional features that are not available on the actual instruments. In the following sections, the installation procedures of the software package will be described. The menu, functions and built-in device categories of the Software VNA will be introduced. The use of the integrated Circuit Simulator in the software package for circuit simulation will also be illustrated. The reference source impedance or port impedance of the Software VNA is set to $Z_{0,\text{ref}} = 50\,\Omega$.

2.1 HOW TO INSTALL

The Software VNA can be run in Microsoft Windows 98, 2000 and XP. To install the software package, run setup.exe in the installation CD. The requirement is that the Software VNA is installed in a file holder called vna in drive c. When the installation is complete, a short cut to the Software VNA can be made and placed on the desktop. The Examples holder in the CD containing all example files can be copied to c:\vna. These examples will be discussed in Chapter 5.

When the Software VNA, i.e. vna.exe, is run on the acceptance of copyright terms, a screen as shown in Figure 2.1 will be displayed.

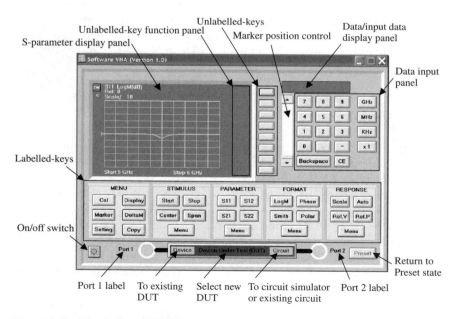

Figure 2.1 The Software VNA

2.2   THE SOFTWARE VNA

The Software VNA shown in Figure 2.1 has a number of features including:

(1) a labelled-key panel;
(2) an unlabelled-key panel;
(3) an unlabelled-key function display panel;
(4) a S-parameter display panel;
(5) a data input panel;
(6) a subdisplay panel;
(7) a marker position control;
(8) a Preset key;
(9) two ports: Port 1 and Port 2;
(10) two connection cables;
(11) a Device Under Test (DUT) key;
(12) a Device short cut key;
(13) a Circuit key; and
(14) a turn-off switch.

The labelled-key panel consists of a STIMULUS function block, a PARAMETER function block, a FORMAT function block, a RESPONSE function block and a MENU block. Each block has five or six keys. Each key has its defined function, which will be described in Section 2.3.

The unlabelled-key panel has eight keys. The function of each key is specified by and can be obtained by clicking on a function key in the MENU function block or Menu labelled keys in other function blocks. The functions are displayed on the unlabelled-key function display panel to the left of the key. Each function can only be valid when the corresponding unlabelled key is clicked. The unlabelled-key functions will be described throughout Sections 2.4 and 2.7 and summarised in Section 2.8.

The S-parameter display panel displays the selected S-parameter across the specified frequency range in a selected format and response specification. The division of the display panel is shown in Figure 2.2 for a typical display. The background and grid colours cannot be changed. However, the colour of the data trace can be changed among blue, red and yellow by double-clicking the display panel. The Software VNA is always calibrated.

The data input panel consists of 0–9 digit keys, a minus '–' sign key, a decimal point '.' key, a backspace key, a clear-entry CE key and four unit keys. Data inputs can be made using these keys. The unit keys, GHz, MHz and KHz, need to be used for frequency input, and x1 key for frequency or other inputs. The input data will be displayed on the subdisplay panel. An input can only be valid when a unit key is functioned. Once a selection which permits data input is made, the subdisplay panel also displays the current value before a new input is made. The subdisplay panel may occasionally display a message.

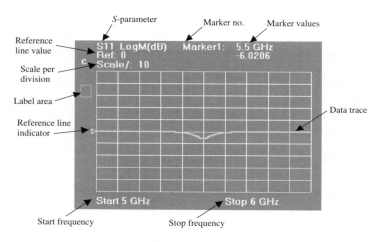

Figure 2.2   The *S*-parameter display panel

The marker position control can be used to change the position of an active marker, which will be discussed in Section 2.7 in detail. The frequency of the active marker will be displayed on the subdisplay panel.

When the Software VNA is started, the switch light is green and a number of default settings are made. A default set of data are also loaded, describing the *S*-parameters of a directly coupled two-port RLC series resonator with a resonance frequency of 5.5 GHz. The settings, such as display format and scale, may be changed as the Software VNA is used. The function of the Preset key is to return the state of the Software VNA to its default settings, which will be described in Section 2.9. The Software VNA can be turned off by clicking on the on/off switch.

The Software VNA provides two ports for one- or two-port measurements and two connection cables for connecting the ports to the DUT. A one-port DUT or network is always taken to be connected to the Port 1 of the Software VNA. Hence S_{11} would be the only valid response of the network in this case. The DUT key gives the user an access to a selection of 35 device categories. From these 35 device categories, an unlimited number of specific devices can be defined. The properties of each device can then be studied. This will be discussed in Section 2.10. The defined devices can also be connected to form a circuit using the integrated Circuit Simulator. This will be described in Section 2.11.

2.3   STIMULUS FUNCTIONS

The STIMULUS functions can be used to set the start and stop frequencies, or centre frequency and frequency span, and the number of frequency points. To set the start frequency, click on the Start key on the STIMULUS

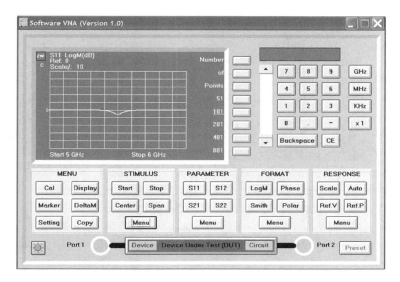

Figure 2.3   STIMULUS menu function display

function block. The current start frequency will be displayed on the subdisplay panel. The input can be made using the keys on the data input panel. Similar steps can be followed to set the stop frequency, centre frequency or frequency span using Stop, Centre and Span keys, respectively, on the same block. The lower and upper limits of the start and stop frequencies are 1 Hz and 1000 GHz, respectively. To set the number of frequency points for the displayed S-parameters, click on Menu on the STIMULUS block. This would lead to a menu shown on the unlabelled-key function display panel, with the current number of points underlined, as shown in Figure 2.3. The number of points can be set by clicking on the corresponding unlabelled key to the right of the display. When the start frequency, stop frequency or the number of points is changed, the responses will be recalculated automatically.

2.4   PARAMETER FUNCTIONS

The PARAMETER functions can be used to select one of the four S-parameters, S_{11}, S_{12}, S_{21} and S_{22}, for display. To select the S-parameter, click on the corresponding key, i.e. S11, S12, S21 or S22, on the PARAMETER function block. Alternatively, click on the Menu key on the same block, which will display S-parameter selections on the unlabelled-key function display panel with the current S-parameter on display underlined, as shown in Figure 2.4. The S-parameter can then be selected by clicking on the corresponding unlabelled key to the right of the display.

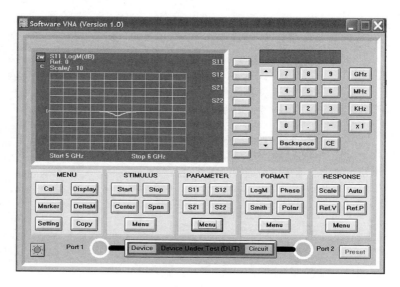

Figure 2.4   PARAMETER menu function display

2.5   FORMAT FUNCTIONS

The FORMAT functions can be used to select one of the following formats for the S-parameter display:

- log-magnitude in dB;
- phase in degrees;
- Smith chart;
- polar chart;
- linear magnitude;
- VSWR; and
- two-port chart.

Examples of log-magnitude, phase, Smith chart, polar chart and two-port chart displays are shown in Figure 2.5(a)–(e).

The selection of log-magnitude in dB, phase in degrees, Smith chart and polar chart can be made by clicking LogM, Phase, Smith and Polar keys, respectively. The Smith chart is available only when S_{11} or S_{22} data are displayed. Other selections can be made by clicking on the Menu key on the FORMAT function block. This would lead to a display depending on whether the DUT is a one- or two-port network. The displays are shown in Figure 2.6(a) and (b), respectively, for a one-port DUT and a two-port DUT. The linear magnitude format, i.e. LinM, is available to both one- and two-port DUTs. However, the VSWR is available only to a one-port DUT. The formats can be selected by clicking on the corresponding unlabelled key to the right of the display. When the Smith chart format is selected, the

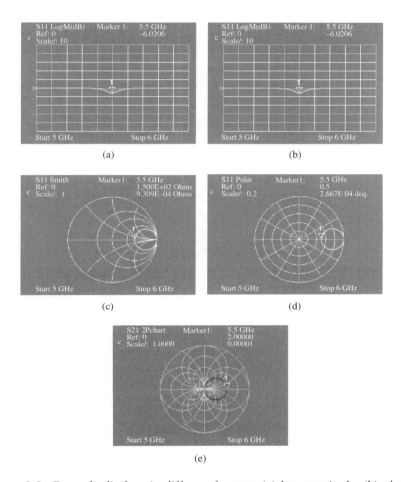

Figure 2.5 Example displays in different formats: (a) log-magnitude, (b) phase, (c) Smith chart, (d) polar chart, (e) two-port chart

choice of impedance or admittance display of active marker values can be made by clicking on the unlabelled key to the right of ShowZ or ShowY.

The two-port chart for single impedance/admittance network, i.e. 2PChart, is available only to a two-port DUT, but for all four types of S-parameters, i.e. S_{11}, S_{12}, S_{21} and S_{22}. The two-port chart has been described in Chapter 1.

2.6 RESPONSE FUNCTIONS

The RESPONSE functions can be used to set scale per division of the display, reference line position, reference line value, electrical delay and phase offset and for autoscale.

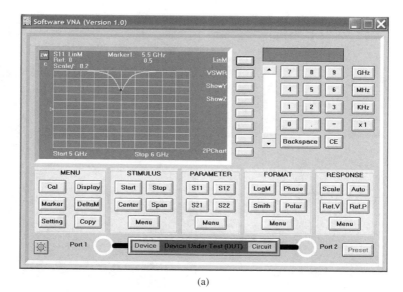

(a)

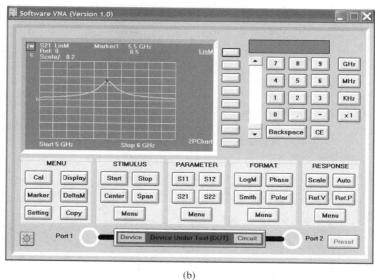

(b)

Figure 2.6 FORMAT menu display for (a) one-port DUT and (b) two-port DUT

To set the scale per division, click on the Scale key on the RESPONSE function block. The current scale per division will be displayed on the subdisplay panel. The input can be made using the keys on the data input panel. The scale per division must be positive. For Smith chart display, the scale per division is always set to 1.

The reference line position and reference line value functions, i.e. Ref.P and Ref.V, are only functional for Cartesian displays. To set the reference line position, click on the Ref.P key on the RESPONSE function block. The current position will then be displayed on the subdisplay panel. The input can be made using the keys on the data input panel. The input value must be between 0 and 10. Otherwise a default value may be set according to the input. To set the reference line value, click on the Ref.V key on the RESPONSE function block. The current value will then be displayed on the subdisplay panel. The input can be made using the keys on the data input panel. The input value may be positive, negative or zero. For non-Cartesian displays, the reference line value is always set to 0, which is generally regarded to be at the centre of the display where the amplitude of the S-parameter is zero.

To set the electrical delay, click on Menu on the RESPONSE block. This would lead to a display on the unlabelled-key function display panel, with the current active function of the menu underlined, as shown in Figure 2.7. Click on the corresponding unlabelled key to the right of E_delay. The current delay length in metres will be displayed on the subdisplay panel. The input can be made using the keys on the data input panel. The delay length represents the total length of two identical 50 Ω air lines connected to both sides of the two-port network. For a one-port network, the delay length is twice the air line length connected to the network due to the delay in signal transmission and reflection. The electrical delay will result in a frequency-dependent phase shifting. When the delay length is not zero, a label D will be displayed on the label area of the S-parameter display panel.

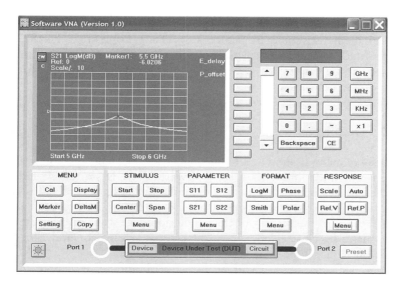

Figure 2.7 RESPONSE menu display

Similarly to set the phase offset, click on Menu on the RESPONSE block. This would again lead to a display on the unlabelled-key function display panel, with the current active function of the menu underlined, as shown in Figure 2.7. Click on the corresponding unlabelled key to the right of P_offset. The current phase offset value in degrees will be displayed on the subdisplay panel. The input can be made using the keys on the data input panel. The unit will be in degrees as the x1 key is clicked. Unlike the electrical delay, the phase offset is frequency independent. The phase response at any frequency point will be offset by the same phase angle. When the phase offset value is not zero, a label O will be displayed in the label area of the S-parameter display panel.

The autoscale function is valid only in Cartesian displays. To use the function, click on the Auto key on the RESPONSE block. This would ensure that the results for all the frequency points are displayed within the grids.

2.7   MENU BLOCK

The MENU block consists of six keys: Cal, Display, Marker, DeltaM, Setting and Copy, which can be used to bring up six different lists of menu displays on the unlabelled-key function display panel. Each menu offers a number of functions as described below, and it can be accessed by clicking on the corresponding key. Each function of the menu can again be activated by clicking on the corresponding unlabelled key to the right of the menu list.

2.7.1   Cal

The Cal menu is shown in Figure 2.8. The 1_PortS11, 1_PortS22 and 2_Ports functions give information on how a commercial vector network analyser can be calibrated using a common, but basic technique for one- and two-port networks.

2.7.2   Display

The Display menu is shown in Figure 2.9. The CalcData function is always selected and underlined when the Software VNA is started. It indicates that the data displayed are a set of simulation data, i.e. calculated data, rather than measured data or data from memory. This function can also be used to return to the display of the simulation data when another function on the menu list is selected. Once CalcData is selected or underlined, the simulation data will become active and be displayed.

The MeasData function can be used to obtain the measured data of S-parameters on a Hewlett-Packard vector network analyser and display the data on the Software VNA. If the DUT on the instrument is a one-port network, the label Port 1 would need to be clicked before MeasData

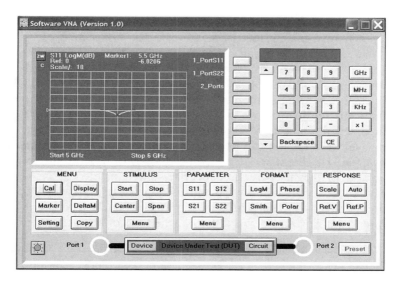

Figure 2.8   Cal menu display

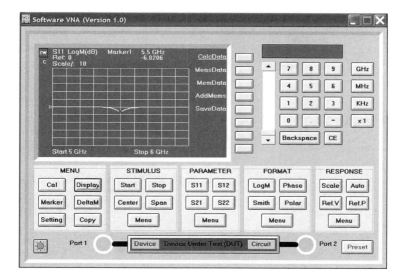

Figure 2.9   Display menu display

function is used, to ensure that only the S_{11} data are transferred, and the Software VNA takes the DUT as a one-port device. Similarly if the DUT on the actual instrument is a two-port network, the label Port 2 would need to be clicked before MeasData function is used, to ensure that all S_{11}, S_{12}, S_{21} and S_{22} data are transferred, and the Software VNA takes the DUT as a two-port device. The number of frequency points will also be read from the instrument, with a maximum number of 801 points.

The MeasData function requires a GPIB PCII or a compatible card produced by National Instruments to be operational. The GPIB card needs to be configured as GPIB0. The address of the Hewlett-Packard vector network analyser needs to be set to 16. When the data transfer cannot be carried out, an error message, GPIB Error, will be displayed on the subdisplay panel. The instrument needs to be calibrated to give meaningful data. Once the data transfer from the instrument is successful, the transferred data, i.e. the measured data, become active and will be displayed.

The SaveData function can be used to save the active S-parameter data in a text file. The file name is preferably chosen to have an affix of .dat. An example of the file name is data1.dat. The data file can be saved in any created file holder for easy reference. For a one-port network, only S_{11} data will be saved as the network is taken to be connected to Port 1 of the Software VNA. For a two-port network, the data of all S-parameters, i.e.

Two-Port	Type of network: one-port or two-port
5000000000	Start frequency in Hz
6000000000	Stop frequency in Hz
101	Number of frequency points
S11amp(dB),S11pha(deg)	Representation of the first line,
S12amp(dB),S12pha(deg)	second line,
S21amp(dB),S21pha(deg)	third line, and
S22amp(dB),S22pha(deg)	fourth line of data below
−3.911667E-02,−3.129202	The first line of data for the first frequency point
−25.24498, 83.7227	The second line of data for the first frequency point
−25.24498, 83.7227	The third line of data for the first frequency point
−3.911667E-02,−3.129202	The fourth line of data for the first frequency point
−4.071709E-02,−3.192142	Repeating above four lines for the second frequency point
−25.07164, 83.59565	
−25.07164, 83.59565	
−4.071709E-02,−3.192142	
. . .	and so on
. . .	
. . .	
−3.911667E-02,3.129202	until the last four lines for the last frequency point
−25.24498,−83.7227	
−25.24498,−83.7227	
−3.911667E-02,3.129202	

S_{11}, S_{12}, S_{21} and S_{22}, will be saved. The data saved are arranged in a format as commented below for a two-port network:

Once a set of data are saved using SaveData, the MemData function can be used, when needed, to load the same set of data onto the Software VNA. The data file to be loaded needs to be selected from the Open Window provided. A change of file holder may be required if the data file happens to be in another file holder. When this function is used, the loaded memory data will become active, replacing simulated data or measured data. A label M will also be displayed in the label area of the S-parameter display panel.

The AddMems function can also be used to load the saved data, i.e. the memory data. However, using this function, the trace of the memory data will be displayed as an addition to the current display. The memory data are not active. Hence any change in the display parameter or format may result in the trace of the memory data to be cleared. The advantage of this function to MemData is that it enables a number of sets of data to be shown and compared on the same graphical display.

Normally the memory data are the saved data from the Software VNA. However, data from other sources may also be displayed using the Software VNA so long as the data are saved in a text file and arranged in the same format as described earlier.

The PrintData function can be used to print the data on display, which can be the simulation data, measured data or data from memory. However, each time the function is used, only the S-parameter data on display are printed on the printer. To print all four S-parameters, each S-parameter has to be selected individually before using the function. The data printed are arranged in the following format, with xx representing a number:

Frequency Response of a One/Two-Port Device
Start Frequency = xx GHz
Stop Frequency = xx GHz
Number of Points = xx
Frequency (GHz)   Sxx_Amp(dB)   Sxx_Pha(deg.)
. . . data lines. . .

2.7.3   Marker

The Marker menu is shown in Figure 2.10. There are five markers in total. Each marker is colour coded as given in Table 2.1. Once a marker is selected, the marker will become active and the values at this active marker position will be displayed on the S-parameter display panel. The values displayed, however, depend on the format of display selected, which is given in Table 2.2.

The position of each marker can be controlled, when it is active, by using the marker position control. Clicking on the arrows would result in the position to change up/down by one frequency point, and the white area

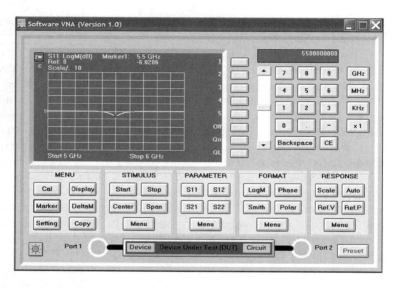

Figure 2.10 Marker menu display

Table 2.1 Colour code of markers

Marker no.	Colour
1	Red
2	Green
3	Blue
4	Yellow
5	Pink

Table 2.2 Marker values at each display format

Format	Values displayed
LogM	Frequency and amplitude of the selected S-parameter in dB
Phase	Frequency and phase of the selected S-parameter in degrees
Smith	Frequency, real and imaginary parts of the impedance or admittance
Polar	Frequency, linear amplitude and phase in degrees of the selected S-parameter
LinM	Frequency and linear amplitude of the selected S-parameter
VSWR	Frequency and VSWR value
2PChart	Frequency, A and B values as defined in Section 1.4.3

to change up/down by 10 points. The control bar can also be dragged to position the marker. The bottom of the marker position control corresponds to the first frequency point and the top to the last frequency point. The total number of frequency points is shown as underlined in the STIMULUS menu. When a marker become inactive, it will still be displayed. All markers can be turned off using the Off function.

The Q_0 and Q_L functions in the menu can be used, respectively, to measure the unloaded and loaded quality factors of a resonator. The measurement, however, depends on whether the device is either one-port or two-port coupled. For a one-port device, the quality factors are measured using the S_{11} response. For a two-port device, they are, however, measured using the S_{21} data assuming the resonator has a bandpass characteristic, even when the S_{11} or S_{22} response is displayed. If this assumption is not met, the values of Q_0 and Q_L are not meaningful. The measured quality factors will be displayed on the subdisplay panel. The Q-factors may be more accurately measured with a smaller frequency span.

2.7.4 DeltaM

The DeltaM menu, as shown in Figure 2.11, provides a list of five markers which can be chosen as a reference or delta marker in Cartesian displays. Once selected, the active marker values displayed in the S-parameter display panel will be referred to the values at the reference maker. A label will also be displayed to indicate the use of the delta marker. For example, Marker 1-2 indicates that the displayed values at Marker 1 (red) position are the

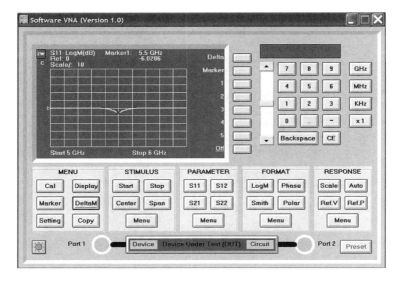

Figure 2.11 DeltaM menu display

increments from the values at Marker 2 (green). Only one delta marker can be selected at a time. The delta marker can be cancelled by using the Off function.

2.7.5 Setting

The Setting menu is shown in Figure 2.12. The Software VNA provides three memories of display format, response and other settings, i.e. Saves 1, 2 and 3, which can be recalled later using the recall functions, Recall 1, 2 and 3, respectively. The information saved includes

device information: one- or two-port;
scales;
reference values;
reference position;
electrical delay;
phase offset;
active marker;
delta marker;
marker positions;
markers: on or off;
S-parameter;
display format;
last function used in each menu;
impedance or admittance marker value display for Smith chart;
trace colour.

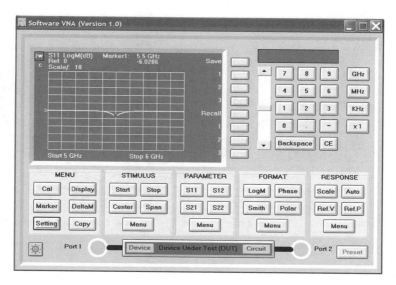

Figure 2.12   Setting menu display

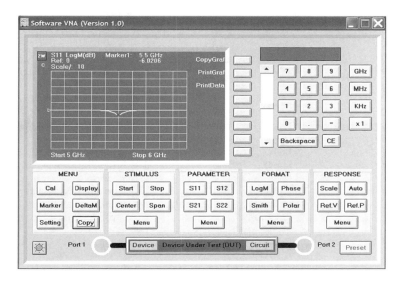

Figure 2.13 Copy menu display

2.7.6 Copy

The Copy menu is shown in Figure 2.13. There are two functions in this menu. The CopyGraf function can be used to copy the S-parameter display to the ClipBoard which could be pasted to other word processor or graphical packages, such as Microsoft Word, Paint or CorelDraw. Using this function, the active marker and delta marker in Cartesian displays will be copied and shown in triangles with a marker number. Inactive markers and delta marker on non-Cartesian displays will not be copied. If all information displayed as seen on the screen need to be copied to the ClipBoard, the Print Screen function key on the keyboard can be used. The information saved on the ClipBoard can then be pasted to other software packages.

The PrintGraf function can be used to print the S-parameter display to a default printer. Again the active and delta markers in Cartesian displays will be printed and shown in triangles with marker numbers. Inactive markers on non-Cartesian displays will not be printed.

2.8 SUMMARY OF UNLABELLED-KEY FUNCTIONS

The unlabelled-key functions described in Sections 2.3–2.7 are summarised in Table 2.3 for quick reference.

Table 2.3   Summary of unlabelled-key functions

Menu	Function name	Function
STIMULUS menu	51, 101, 201, 401, 801	To set the number of points indicated
PARAMETER menu	S11, S12, S21, S22	To set indicated S-parameter for display
FORMAT menu	LinM	To set linear magnitude display of S-parameter
	VSWR	To display VSWR
	ShowY	To show admittance values at active marker position
	ShowZ	To show impedance values at active marker position
	2PChart	To display S-parameter in two-port chart
RESPONSE menu	E_delay	To set electrical delay in metres
	P_offset	To set phase offset in degrees
Cal	1_PortS11, 1_PortS22, 2_Ports	To show how an actual vector network analyser can be calibrated using a common technique
Display	CalcData	To display the calculated/simulated data
	MeasData	To access the data on the actual HP VNA and display
	MemData	To display saved/memory data
	AddMems	To add trace of saved/memory data to the display
	SaveData	To save the data currently on display
	PrintData	To print the data on display
Marker	1, 2, 3, 4, 5	To select or turn on a marker
DeltaM	1, 2, 3, 4, 5	To set a marker as a delta/reference marker
Setting	Save: 1, 2, 3	To save display settings
	Recall: 1, 2, 3	To recall display settings
Copy	CopyGraf	To copy the S-parameter display to the ClipBoard
	PrintGraf	To print the S-parameter display

2.9   PRESET

When the Preset key is clicked, the following default settings are made:

Start frequency = 5 GHz
Stop frequency = 6 GHz
Format: set to LogM
Scale = 10 for LogM, 40 for Phase, 1 for Smith chart, 0.2 for Polar chart,

1 for two-port chart, 0.2 for LinM and 5 for VSWR
Reference value $= 0$
Reference position $= 5$
Electrical delay $= 0$
Phase offset $= 0$
Markers: set off for all markers
Delta markers: set off for all markers
Marker positions: set to the first point for all markers
Number of point $= 101$
Display trace colour set to yellow
Displayed data set to the S-parameter data of a resonator with $f_0 = 5.5$ GHz.

2.10 DEVICE UNDER TEST

The DUT key can be used to define a device using the built-in device builders, to select a device if it has already been built or to get access to the Circuit Simulator. When the key is clicked, a window with a title of Device Under Test as shown in Figure 2.14 will be displayed. There are 35 device builders in the Software VNA. They are listed in Table 2.4. Their corresponding icons, listed in Table 2.4, are also shown in the window. Details of the device builders and their models will be given in Chapter 3. The Device Under Test window has two menus, i.e. Device and Circuit. The device builders can be accessed through the Device menu or the icons in the

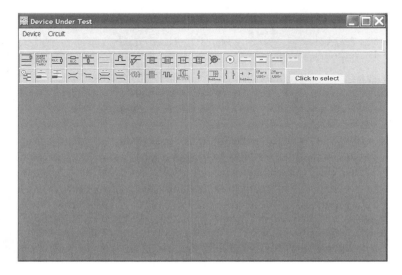

Figure 2.14 Device Under Test (DUT) display

Table 2.4   List of device builders and their icons

No.	Device builder group	Device builder	Icon
0	Demo	Lossless Transmission Line	
1	Standards	One- and Two-Port Standards	
2	Discrete RLC Components	One-Port Impedance Load	
3		Two-Port Series Impedance	
4		Two-Port Shunt Admittance	
5	Transmission Line	General Transmission Line	
6	Transmission Line Components	Two-Port Serial Transmission Line Stub	
7		Two-Port Parallel Transmission Line Stub	
8	Ideal Two-Port Components	Attenuator/Gain Block	
9		$1:N$ or $N:1$ Transformer	
10		Isolator	
11		Gyrator	
12		Circulator	
13	Physical Transmission Lines	Coaxial Line	
14		Microstrip Line	
15		Stripline	
16		Coplanar Waveguide	
17		Coplanar Strip	
18	Physical Line Discontinuities	Coaxial Line Discontinuities	
19		Microstrip Discontinuities	
20		Stripline Discontinuities	
21	General Coupled Lines	Four-Port Coupled Lines	
22		Two-Port Coupled Lines	
23	Physical Coupled Lines	Four-Port Coupled Microstrip Lines	
24		Two-Port Coupled Microstrip Lines	
25	Lumped Elements	Inductors	
26		Capacitors	
27		Resistors	
28	Active Devices	Active Devices	

29		Dipole Antenna	
30		Resonant Antenna	
31	Antennas	Transmission between Dipole Antennas	
32		Transmission between Resonant Antennas	
33	User-Defined *S*-Parameters	One-Port Device	
34		Two-Port Device	

Software VNA window. Similarly, the Circuit Simulator can be accessed from the Circuit menu.

Once a device is defined or selected by the user and the results of *S*-parameters are displayed in the Software VNA, the device can be accessed directly from the Device key in the Software VNA window. Otherwise the default Demo: Lossless Transmission Line device builder is shown when the Device key is clicked. The Circuit Simulator can also be accessed through the Circuit key in the Software VNA window. If a circuit has already been created, the circuit will be displayed. The Device menu and Circuit menu are further described below.

2.10.1 Device

The details of the Device menu are shown in Figure 2.15. It includes the device builders listed in Table 2.4 and two other functions: Open and Cancel.

The device builders and models will be described in Chapter 3 in detail. The selection of a device builder and definition of component parameters are illustrated here using a lossless transmission line demo. The Demo: Lossless Transmission Line device builder can be accessed by selecting the Demo: Lossless Transmission Line from the Device menu. This leads to the display of the window with a title Device Under Test – [Demo: Lossless Transmission Line], as shown in Figure 2.16. A selection of parameters can then be made. The transmission line can be defined as either a one-port or a two-port device. If it is defined as a two-port device, inputs of the characteristic impedance of the transmission line, phase velocity and line length are required. The inputs can be made on the input slots provided by changing the default or existing values. If the transmission line is defined as a one-port device, further inputs of the parameters for the termination load are required. The load can be selected to be a short-circuit or open-circuit load or a load with RLC connected in parallel or in series. In case the load selected is an RLC connected in parallel or in series, inputs of *R*, *L* and *C* parameters are required. Once all the inputs are made, the

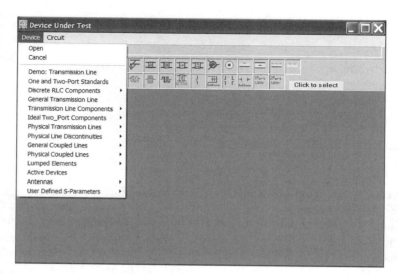

Figure 2.15   Device menu

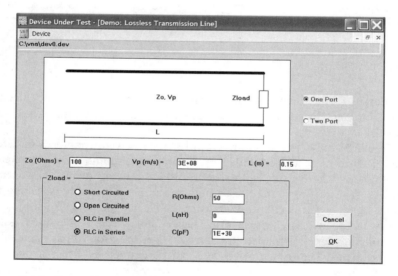

Figure 2.16   Demo: Lossless Transmission Line

defined device can be saved to a file with an affix of .dev, by using the Save or Save As function in the new Device menu. Similarly the defined device can be printed using the Print function in the new Device menu. To define the device as the DUT of the Software VNA, click on OK key to proceed. The selected *S*-parameters will be displayed. Otherwise click on Cancel key to return to the Device Under Test window shown in Figure 2.14.

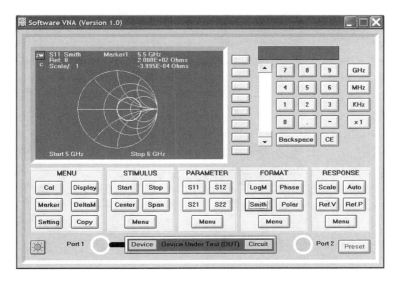

Figure 2.17   The S_{11} display in Smith chart format of the defined device in Figure 2.16

With the parameters as defined in Figure 2.16, i.e. a 15 cm, 100 Ω air line terminated by a 50 Ω load, the display of S_{11} parameters of the defined device in Smith chart format is shown in Figure 2.17. The same procedures can be used to modify the input parameters, if saved previously, and save the device as a new device. The procedures can be repeatedly used to select other device builders, input device parameters and save the defined devices.

The Software VNA has 35 default devices, one for each device builder. When a device builder is selected for the first time, the default device with a name dev*.dev is loaded. The default inputs can be modified, and the defined device can be saved as a new device. The Software VNA keeps a record of the last device defined for each device builder.

The Open function in the Device menu in the Device Under Test window can be used to open a previously defined device. The Cancel function in the Device menu would return the window display to the previous window.

If the start frequency, stop frequency or the number of points is changed, the responses will be recalculated automatically.

2.10.2   Circuit

The Circuit menu provides two functions, i.e. Run Circuit Simulator and Cancel, as shown in Figure 2.18. The Cancel function can be used to leave the current window and return to the previous window. The Run Circuit Simulator function can be used to start the circuit simulator with a built-in

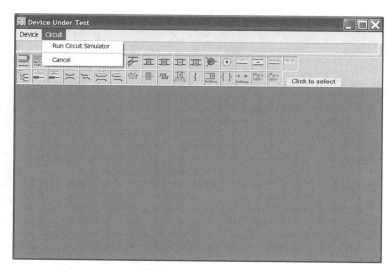

Figure 2.18 Circuit Simulator access menu

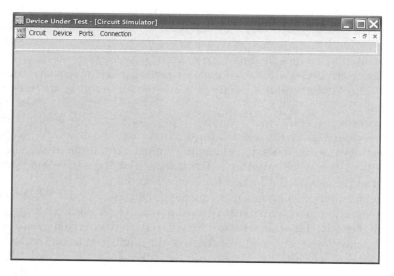

Figure 2.19 Circuit Simulator

circuit assembler. When it is started, a window as shown in Figure 2.19 is displayed. The Circuit Simulator is ready for defining circuit blocks and specifying devices in each block, which is further described in the next section.

2.11   CIRCUIT SIMULATOR

The Circuit Simulator assumes that the circuit to be assembled is composed of a number of blocks. Each block can be a one-port or two-port device defined in the same way as that described in Section 2.10. The maximum number of blocks that can be defined in the Circuit Simulator is 20 and the maximum number of internal ports is 40. The Circuit Simulator, as shown in Figure 2.19, has four menus: Circuit, Device, Ports and Connections.

2.11.1   Circuit Menu

The Circuit menu provides the following functions: New, Open, Show Device List, Save, Save As, Create Holder, Print, Cancel and Run, as shown in Figure 2.20.

The New function can be used to create a new circuit. When it is selected, the user will be prompted with a Create a New Circuit Holder window so that the circuit can be saved into the specified file holder. The circuit can be saved using Save or Save As into a file with an affix .nwk. If Save As is to be used, the Create Holder can be used first to create a new file holder for the new circuit. The saved circuit file can be opened using the Open function. When a circuit is saved or opened, the name of the circuit will be displayed in the circuit name slot below the menus. The Show Device List function can be used to show the list of the devices in circuit. The circuit display can be printed using the Print function.

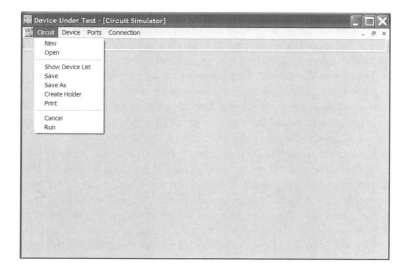

Figure 2.20   Circuit menu

The *S*-parameters of the circuit can be simulated using the Run function. The *S*-parameters will be displayed on the Software VNA. The simulation results can be investigated, and the properties of the circuit can be studied. However, if a circuit is not complete, an error message may appear. The circuit needs to be completed before applying the Run function again. The Cancel function can be used to return to the Software VNA window.

2.11.2   Device Menu

The Device menu provides the following functions: New, Undo, Copy, Delete and Paste, as shown in Figure 2.21(a).

The New function can be used to create a new device for the circuit. When it is selected, a Circuit Block Construction panel will appear, as shown in Figure 2.21(b). The default name of the device can be changed. The Device Type can be selected to be D, T, M, S or O. For devices to be defined using the device builders, D should be chosen. Otherwise T, M, S and O should be chosen for a T-junction, matched load, short load and open load, respectively.

To define a device using a device builder, the icon of the device on the icon list should be dragged and dropped to the icon slot to the right of the Device Type. Alternatively the icon of the device on the icon list can be double-clicked so that the icon of the device will appear on the icon slot mentioned. This icon appearing in the icon slot can then be doubled-clicked to define the device. When the device builder is displayed, the parameters of the device can be inputted. The definition of the device is completed when the OK key is clicked. The device can then be saved with a given device name with affix .dev. For easy management of the device, the device should be saved to the holder created for the circuit. If it is saved in a different holder, a copy of the device will also be saved in the same holder of the circuit. The process is cancelled if the Cancel key is activated. In this case, an icon of the selected device will appear on the icon slot to the right of the Device Type. The device can be redefined using the same procedure. When the device is saved, the device file name will be displayed on the Device File information slot. A note about the device can also be added using the note slot. The device can be added to the circuit by clicking on Add. Otherwise clicking on Cancel will cease the creation of the device.

When a device is added to the circuit, a circuit block with appropriate ports will be shown in the Circuit Display Area. A number in brackets, e.g. (0), will be displayed after the name of the block to indicate the block number as it is created. The position of the circuit block can be moved by dragging and dropping the block. The information of the defined block can be accessed by double-clicking the block. The name of the circuit block will be shown in red. The information and the parameters of the device associated with the block can be changed, and the changes can be saved using the Save Changes key. The Circuit Block Construction panel can be

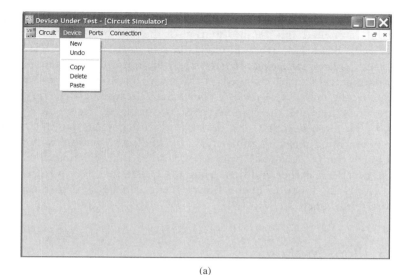

(a)

(b)

Figure 2.21 (a) Device menu and (b) device building

closed by clicking on the Cancel key. This process can also be applied when the circuit contains more blocks.

A circuit block can be copied using the Copy and Paste functions. This is made by first selecting the circuit block to be copied, i.e. double-clicking the block. The name of the block will be shown in red and the Circuit Block Construction panel will appear. When this is done, the Copy and Paste functions in the Device menu can be used to copy the block. A new circuit block will be created and the information of the new block will be shown in the Circuit Block Construction panel. At this stage, the copied device is identical to its original device. However, all the information including the device parameters can be changed. If the device parameters are changed, the name of the new device needs to be changed accordingly. If the device name remains the same, the parameters of its original device would also be changed with the new inputs. If a device is already copied, multiple copies can be made using repeated applications of the Paste function. Similarly a circuit block can be deleted. This can be made by first selecting the circuit block as described earlier and then using the Delete function in the Device menu. The Undo function in the Device menu can be used to revert the Paste or Delete function.

2.11.3 Ports Menu

The Ports menu provides the following functions: Add Port1, Add Port2 and Remove Port2, as shown in Figure 2.22. A one-port circuit needs only to connect to Port 1 of the Software VNA, but a two-port circuit requires connections to both Ports 1 and 2 of the Software VNA. In both cases, the Add Port1 function needs to be used to show Port 1 of the Software VNA. The Add Port2 function is used only for two-port circuits. The Remove Port2 function can be used to remove the display of Port 2 of the Software VNA when required.

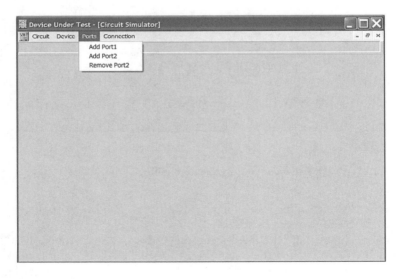

Figure 2.22 Ports menu

2.11.4 Connection Menu

The Connection menu contains the How to Connect function only, as shown in Figure 2.23. It shows the user how the circuit blocks defined can be connected together to form a circuit. A connection can be made between any two ports in two different blocks including Port 1 and Port 2 of the Software VNA. This is done by clicking the port to be connected in one block and the other port to be connected in another block. The connected ports will be shown in red. To disconnect the ports, click one of the connected ports. The disconnected ports will be shown in the default colour. When a port is selected for connection, it will be displayed in red. To deselect the port, click on the port again and it will be displayed in the default colour.

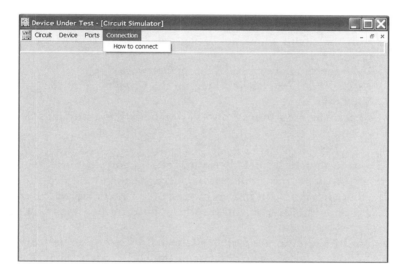

Figure 2.23 Connection menu

2.12 CIRCUIT SIMULATION PROCEDURES AND EXAMPLE

The procedures for the simulation of a circuit are outlined below:

(1) Sketch the circuit to be simulated.
(2) Break the circuit into circuit blocks so that each block can be modelled by a device that can be built using the device builders in the Software VNA.

(3) Identify the input and output ports of the circuit.

(4) Run the Software VNA (if it is not already run).

(5) Specify the start and stop frequencies and the number of frequency points.

(6) Run Circuit Simulator.

(7) Create a new circuit holder using the New function in the Circuit menu.

(8) Create the required devices using the functions in the Device menu, and save the devices in the circuit holder.

(9) Add Port 1 and Port 2 of the Software VNA using the functions in the Ports menu.

(10) Move the positions of the devices, Port 1 and Port 2, as required.

(11) Connect the ports of the circuit blocks to form a circuit.

(12) Connect the input and output ports of the circuit to Port 1 and Port 2 of the Software VNA.

(13) Save the circuit in the circuit holder.

(14) Run the circuit simulation using the Run function in the Circuit menu.

(15) View the simulation results on the Software VNA.

(16) Re-view the circuit by clicking on the Circuit key in the Software VNA window.

(17) Repeat the procedures from (8) to modify the circuit and simulate if required.

However, if a circuit is already saved, it can be opened using the Open function in the Circuit menu. Once it is opened, it can be modified and saved in a different circuit holder created by using the Create Holder function if needed. Procedures (13)–(15) can then be applied to the simulation of the circuit opened or modified.

The above procedures are further illustrated using the circuit of a 3 dB Wilkinson power divider. The design and simulation of the divider will be further discussed in Sections 4.4.1 and 5.5.1, respectively. Figure 2.24(a) shows the sketched circuit of the 3 dB Wilkinson power divider with a central frequency $f_0 = 3$ GHz. On inspection, the circuit is made of seven blocks: two transmission lines of characteristic impedance $Z_0 = \sqrt{2}Z_{0,\text{ref}} = 70.7\,\Omega$ where $Z_{0,\text{ref}} = 50\Omega$ and length of 25 mm with $v_p = c$, one isolation resistor of $R = 100\,\Omega$, one matched load and three T-junctions. The divider is a three-port circuit. But for simulation, only one of the ports can be taken as the input port and one of the remaining ports as the output port. The final remaining port needs to be terminated by a matched load. Hence Port 1 of the divider is taken as the input port and Port 2 of the divider as the output port. Port 3 of the divider is terminated by a matched load, as shown in Figure 2.24(b).

Once the circuit blocks are identified, run the Software VNA. For the central frequency of the divider of 3 GHz, set the start frequency and stop

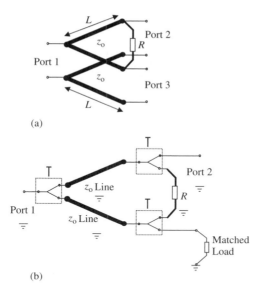

(a)

(b)

Figure 2.24   3 dB Wilkinson power divider: (a) divider circuit and (b) circuit for simulation

frequency to 1 and 5 GHz, respectively. Set the number of frequency points to 101. Run the Circuit Simulator by clicking on the Circuit key in the Software VNA window.

Click on the Circuit menu. Select New. Navigate the file holder directories to create a circuit holder, e.g. under C:\vna\. Give a name to the circuit holder, e.g. Example. Click on the Create key to create the circuit holder, i.e. C:\vna\Example for the example.

Click on the Device menu and select New. The Circuit Block Construction panel will then appear. Select the Device Type to be T. Change the default name of the device from Block0 to T. Add a note if needed. Click on the Add key to create the first block T(0). The number after the name of the block in brackets indicates the block number. The T-junction has three ports.

Click on the Device menu and select New. The Circuit Block Construction panel will again appear. Select the Device Type to be D, which is the default selection of Device Type. Change the default name of the device from Block1 to Line. Drag the General Transmission Line icon on the icon list and drop it to the icon slot to the right of the Device Type. Input $Z_0 = 70.7\,\Omega$, $\alpha = 0$, Frequency Dependence to the power of 0.5, $v_p/300000000 = 1$, $L = 0.025\,m$ accordingly. Click on the OK key. Save the device in C:\vna\Example circuit holder as line.dev. Add a note '70.7 Ohms T-line' to the device for quick reference. Click on the Add key to create the second block Line(1). The transmission line has two ports.

Double-click the Line(1) block. Click on Device menu. Select Copy. Click on Device menu again. Select Paste. A new block Line(2) is created with the same properties as Line(1). Click the Cancel key to close the Circuit Block Construction panel. Drag the block to its appropriate position.

Double-click the T(0) block. Click on Device menu. Select Copy. Click on Device menu again. Select Paste. A new T-junction block T(3) is created. Drag the block to its appropriate position. Click on Device menu. Select Paste to create the T-junction block T(4). Drag the block to its appropriate position. Click the Cancel key to close the Circuit Block Construction panel.

Click on the Device menu and select New. Select the Device Type to be D on the Circuit Block Construction panel. Change the default name of the device from Block5 to R. Drag the Two-Port Series Impedance icon on the icon list and drop it to the icon slot to the right of the Device Type. Input $R = 100\,\Omega$, $L = 0\,nH$, $C = $ No for the components in series and $R = 0\,\Omega$, $L = 0$, $C = $ No for the components in parallel accordingly. Click on the OK key. Save the device in C:\vna\Example circuit holder as R.dev. Add a note '100 Ohms resistor' to the device. Click on the Add key to create the sixth block R(5). Drag the block to its appropriate position. The resistor device has two ports.

Click on the Device menu and select New. On the Circuit Block Construction panel, select the Device Type to be M. Change the default name of the device from Block6 to Matched. Add a note if needed. Click on the Add key to create the seventh block Matched(6). Drag the block to its appropriate position. The matched load has one port.

Click on the Ports menu and select Add Port1. Click on the Ports menu again and select Add Port2. Drag Port 1 and Port 2 to their appropriate positions.

All the circuit blocks have now been created. Connect Port 1 of the Software VNA to Port 1 of T(0). Connect Port 2 of T(0) to Port 1 of Line(1). Connect Port 3 of T(0) to Port 1 of Line(2). Connect Port 2 of Line(1) to Port 1 of T(3). Connect Port 2 of Line(2) to Port 1 of T(4). Connect Port 2 of T(3) to Port 1 of R(5). Connect Port 3 of T(4) to Port 2 of R(2). Connect Port 2 of T(4) to Port 1 of Matched(6). Connect Port 3 of T(3) to Port 2 of the Software VNA. The circuit is now complete, as shown in Figure 2.25. Click on the Circuit menu and select the Save As function. Save the circuit in the circuit holder C:\vna\Example as Example.nwk. Click on the Circuit menu and select the Show Device List function to view the list of devices in the circuit if needed.

Click on the Circuit menu and select the Run function to simulate. The S-parameter responses of the circuit will be displayed in the Software VNA, as shown in Figure 2.26(a) and (b).

Click on the Circuit key on the Software VNA to return to the Circuit Simulator window.

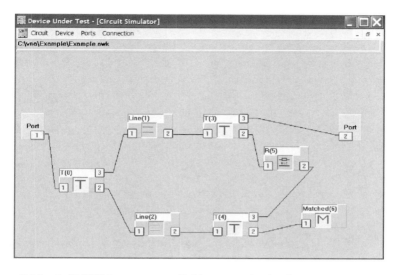

Figure 2.25   3 dB Wilkinson power divider example circuit

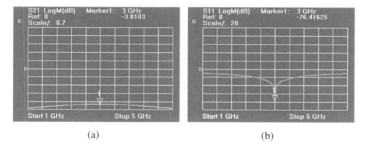

(a)   (b)

Figure 2.26 S-parameters of the 3 dB Wilkinson power divider: (a) S_{21} response, (b) S_{11} response

If the start frequency, stop frequency or the number of points is changed, the responses will be recalculated automatically.

The same example has been included in C:\vna\also as Example.nwk together with attached devices Line.dev and R.dev.

3

Device Builders and Models

ABSTRACT

The Software VNA has 35 device builders, from which an unlimited number of devices can be defined. The devices built can be used for simulation studies either as a single DUT or as a part of a circuit to be assembled using the Circuit Simulator. In this chapter, these device builders are described in detail. The equations used for the modelling of devices are provided.

KEYWORDS

Microwave devices, Device modelling equations, Discrete components, Ideal components, Transmission lines, Active components, Transmission line discontinuities, Coupled transmission lines, Lumped elements, Antennas, User-defined devices.

The Software VNA described in Chapter 2 has 35 device builders, which are listed again in Table 3.1. An unlimited number of devices can be defined using these device builders, which can then be used for simulation studies either as a single DUT or as part of a circuit assembled using the Circuit Simulator. In this chapter, these device builders will be described in detail. A list of equations used for device modelling will be given. For the S-parameter measurements, the reference port impedance is always assumed to be $Z_{0,\mathrm{ref}} = 50\,\Omega$ or the reference port admittance to be $Y_{0,\mathrm{ref}} = 1/Z_{0,\mathrm{ref}}$. In the case of one-port device, it is assumed that the DUT is connected to Port 1 of the Software VNA, and Port 2 of the Software VNA is left

Software VNA and Microwave Network Design and Characterisation   Zhipeng Wu
© 2007 John Wiley & Sons, Ltd

Table 3.1   List of device builders

No.	Device builder group	Device builder
0	Demo	Lossless Transmission Line
1	Standards	One- and Two-Port Standards
2	Discrete RLC Components	One-Port Impedance Load
3		Two-Port Series Impedance
4		Two-Port Shunt Admittance
5	Transmission Line	General Transmission Line
6	Transmission Line Components	Two-Port Serial Transmission Line Stub
7		Two-Port Parallel Transmission Line Stub
8	Ideal Two-Port Components	Attenuator/Gain Block
9		1:N or N:1 Transformer
10		Isolator
11		Gyrator
12		Circulator
13	Physical Transmission Lines	Coaxial Line
14		Microstrip Line
15		Stripline
16		Coplanar Waveguide
17		Coplanar Strip
18	Physical Line Discontinuities	Coaxial Line Discontinuities
19		Microstrip Discontinuities
20		Stripline Discontinuities
21	General Coupled Lines	Four-Port Coupled Lines
22		Two-Port Coupled Lines
23	Physical Coupled Lines	Four-Port Coupled Microstrip Lines
24		Two-Port Coupled Microstrip Lines
25	Lumped Elements	Inductors
26		Capacitors
27		Resistors
28	Active Devices	Active Devices
29	Antennas	Dipole Antenna
30		Resonant Antenna
31		Transmission between Dipole Antennas
32		Transmission between Resonant Antennas
33	User-Defined S-Parameters	One-Port Device
34		Two-Port Device

'open' so that $S_{22} = 1$ and $S_{12} = S_{21} = 0$. For calculations, the zero value is approximated by 10^{-30} in the Software VNA, and an infinite value by 10^{30}. The speed of light in air is taken to be $c = 3 \times 10^8$ m/s.

3.1   LOSSLESS TRANSMISSION LINE

When the Lossless Transmission Line device builder is selected from the Device Under Test (DUT) list, a one-port transmission line as shown in

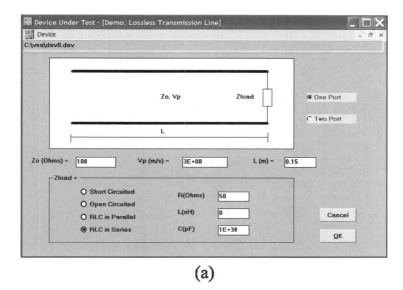

(a)

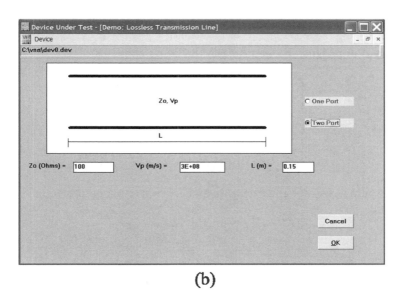

(b)

Figure 3.1 Lossless Transmission Line: (a) One-Port option and (b) Two-Port option

Figure 3.1(a) will be displayed. The characteristic impedance, propagation velocity and length of the lossless transmission line can be assigned. The termination load can be selected to be a short-circuit load, an open-circuit load, a load with RLC connected in parallel or a load with RLC connected in series. The RLC values can be inputted. If a component does not exist, the corresponding input field can be either left blank or assigned an input of No. The 'nonexisting' component will be replaced by a short circuit across the component in the series case and by an open circuit in the parallel case. The Software VNA will assign an appropriate 'zero' or 'infinite' value for the component. If the option of Two-Port is selected, the two-port lossless transmission line is as shown in Figure 3.1(b).

The S-parameter of the one-port lossless transmission line is given by

$$S_{11} = \frac{Z_{in} - Z_{0,ref}}{Z_{in} + Z_{0,ref}}, \tag{3.1}$$

where

$$Z_{in} = Z_0 \frac{Z_L + jZ_0 \tan(\beta L)}{Z_0 + jZ_L \tan(\beta L)} \tag{3.2}$$

and

$$\beta = \frac{\omega}{v_p} = \frac{2\pi f}{v_p}. \tag{3.3}$$

The load impedance Z_L depends on the user's choice, and

$$Z_L = \begin{cases} 0 & \text{short circuit} \\ \infty & \text{open circuit} \\ \left[R + j\left(\omega L - \dfrac{1}{\omega C} \right) \right]^{-1} & \text{RLC in parallel.} \\ \left[R + j\left(\omega L - \dfrac{1}{\omega C} \right) \right] & \text{RLC in series} \end{cases} \tag{3.4}$$

The S-parameters for the two-port lossless transmission line are given by

$$[S] = \frac{\begin{bmatrix} j(Z_0^2 - Z_{0,ref}^2) \sin(\beta L) & 2Z_0 Z_{0,ref} \\ 2Z_0 Z_{0,ref} & j(Z_0^2 - Z_{0,ref}^2) \sin(\beta L) \end{bmatrix}}{2Z_0 Z_{0,ref} \cos(\beta L) + j(Z_0^2 + Z_{0,ref}^2) \sin(\beta L)}. \tag{3.5}$$

3.2   ONE- AND TWO-PORT STANDARDS

Figure 3.2 shows the display of the One- and Two-Port Standards. The standards include Short, Open, Matched Load or Thru. They are directly connected to the ports of the Software VNA with zero transmission line

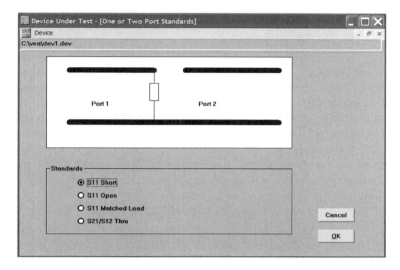

Figure 3.2 One- and Two-Port Standards

lengths. The one-port standards are connected to Port 1 of the Software VNA and the two-port standard to both Port 1 and Port 2 of the Software VNA.

The S-parameters for the standards are tabulated in Table 3.2.

Table 3.2 The S-parameters of one- and two-port standards

Standards	S-parameters
Port 1: Short	$[S] = \begin{bmatrix} -1 & 0 \\ 0 & 1 \end{bmatrix}$
Port 1: Open	$[S] = \begin{bmatrix} 1 & 0 \\ 0 & 1 \end{bmatrix}$
Port 1: Matched Load	$[S] = \begin{bmatrix} 0 & 0 \\ 0 & 1 \end{bmatrix}$
Thru	$[S] = \begin{bmatrix} 0 & 1 \\ 1 & 0 \end{bmatrix}$

Note: Port 2 of the Software VNA is left open for one-port connection to Port 1

3.3   DISCRETE RLC COMPONENTS: ONE-PORT IMPEDANCE LOAD

Figure 3.3 shows the display of the Discrete RLC Components: One-Port Impedance Load device builder. The load consisting of both RLC in series and RLC in parallel is connected to Port 1 of the Software VNA with zero transmission line length. All RLC values can be inputted. If a component does not exist, the corresponding input field can be either left blank or assigned an input of No. The 'nonexisting' component will be replaced by a short circuit in the series case and by an open circuit in the parallel case. The Software VNA will assign an appropriate 'zero' or 'infinite' value accordingly.

The S-parameter of the one-port impedance load is given by

$$S_{11} = \frac{Z_L - Z_{0,\text{ref}}}{Z_L + Z_{0,\text{ref}}}, \tag{3.6}$$

where

$$Z_L = R_s + j\left(\omega L_s - \frac{1}{\omega C_s}\right) + \frac{R_p}{1 + jR_p\left(\omega C_p - \frac{1}{\omega L_p}\right)}. \tag{3.7}$$

The subscripts s and p represent the series and parallel components, respectively.

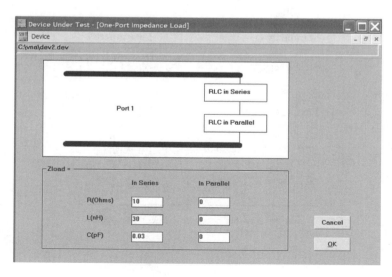

Figure 3.3   Discrete RLC Components: One-Port Impedance Load

3.4   DISCRETE RLC COMPONENTS: TWO-PORT SERIES IMPEDANCE

Figure 3.4 shows the display of the Discrete RLC Components: Two-Port Series Impedance device builder. The impedance consists of R, L, C components connected in series and other three connected in parallel. The impedance is connected to both Port 1 and Port 2 of the Software VNA with zero transmission line lengths. All six RLC values can be inputted. If a component does not exist, the corresponding input field can be either left blank or assigned an input of No. The 'nonexisting' component will be replaced by a short circuit in the series case and by an open circuit in the parallel case. The Software VNA will assign an appropriate 'zero' or 'infinite' value accordingly.

The S-parameters of the two-port series impedance device are given by

$$[S] = \begin{bmatrix} \dfrac{Z}{Z+2Z_{0,\mathrm{ref}}} & \dfrac{2Z_{0,\mathrm{ref}}}{Z+2Z_{0,\mathrm{ref}}} \\[3mm] \dfrac{2Z_{0,\mathrm{ref}}}{Z+2Z_{0,\mathrm{ref}}} & \dfrac{Z}{Z+2Z_{0,\mathrm{ref}}} \end{bmatrix}, \tag{3.8}$$

where

$$Z = R_\mathrm{s} + j\left(\omega L_\mathrm{s} - \frac{1}{\omega C_\mathrm{s}}\right) + \frac{R_\mathrm{p}}{1+jR_\mathrm{p}\left(\omega C_\mathrm{p} - \dfrac{1}{\omega L_\mathrm{p}}\right)}. \tag{3.9}$$

The subscripts s and p represent the series and parallel components, respectively.

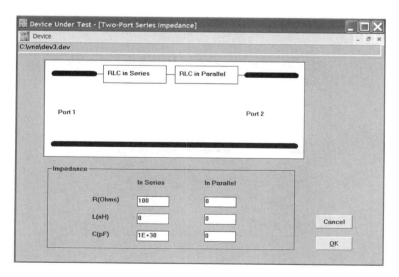

Figure 3.4   Discrete RLC Components: Two-Port Series Impedance

3.5   DISCRETE RLC COMPONENTS: TWO-PORT SHUNT ADMITTANCE

Figure 3.5 shows the display of the discrete RLC components: Two-Port Single Admittance device builder. The admittance consists of R, L, C components connected in series and other three connected in parallel. The shunt admittance is connected to both ports of the Software VNA with zero transmission line lengths. The RLC values can be inputted. If a component does not exist, the corresponding input field can be either left blank or assigned an input of No. The 'nonexisting' component will be replaced by a short circuit in the series case and by an open circuit in the parallel case. The Software VNA will assign an appropriate 'zero' or 'infinite' value accordingly.

The S-parameters of the two-port single admittance device are given by

$$[S] = \begin{bmatrix} \dfrac{-YZ_{0,\text{ref}}}{2 + YZ_{0,\text{ref}}} & \dfrac{2}{2 + YZ_{0,\text{ref}}} \\ \dfrac{2}{2 + YZ_{0,\text{ref}}} & \dfrac{-YZ_{0,\text{ref}}}{2 + YZ_{0,\text{ref}}} \end{bmatrix}, \tag{3.10}$$

where

$$Y = \left[R_\text{s} + \text{j}\left(\omega L_\text{s} - \frac{1}{\omega C_\text{s}} \right) + \frac{R_\text{p}}{1 + \text{j}R_\text{p}\left(\omega C_\text{p} - \dfrac{1}{\omega L_\text{p}} \right)} \right]^{-1}. \tag{3.11}$$

The subscripts s and p represent the series and parallel components, respectively.

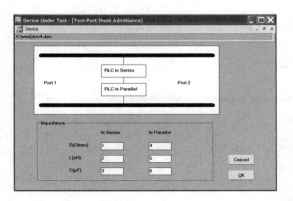

Figure 3.5   Discrete RLC Components: Two-Port Shunt Admittance

3.6   GENERAL TRANSMISSION LINE

Figure 3.6 shows the display of the General Transmission Line device builder. Unlike the lossless transmission line in Section 3.1, the general transmission line can be lossy, i.e. the attenuation constant may have a finite nonzero value. The transmission line is a two-port device and is connected to both ports of the Software VNA. The transmission line parameters including characteristic impedance, attenuation constant at 1 GHz, frequency dependence of the attenuation constant, phase velocity w.r.t. the speed of light in air and length can be inputted.

The attenuation constant is usually frequency and geometry dependent. A value at 1 GHz needs to be obtained for a given transmission line geometry. For a lossless transmission line, the attenuation constant is 0. For normal conductors, the attenuation constant due to conducting loss increases proportionally to the square root of frequency. The frequency dependence factor of the attenuation constant is 0.5. For superconductors, the factor may be between 1 and 2. The factor may be 1 for attenuation due to dielectric losses.

Both characteristic impedance and phase velocity of a transmission line depend on its geometry and dielectric filling between the conductors. For a pure TEM transmission line, the phase velocity is given by

$$v_\mathrm{p} = \frac{c}{\sqrt{\varepsilon_\mathrm{r}}},\qquad(3.12)$$

where c is the speed of light in air and ε_r the relative dielectric constant of the dielectric filling in the transmission line.

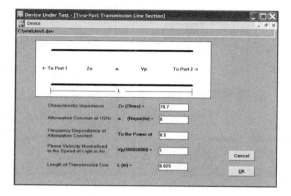

Figure 3.6   General Transmission Line

The S-parameters of the general transmission line are given by

$$[S] = \frac{\begin{bmatrix} (Z_0^2 - Z_{0,\text{ref}}^2)\sinh(\gamma L) & 2Z_0 Z_{0,\text{ref}} \\ 2Z_0 Z_{0,\text{ref}} & (Z_0^2 - Z_{0,\text{ref}}^2)\sinh(\gamma L) \end{bmatrix}}{2Z_0 Z_{0,\text{ref}}\cosh(\gamma L) + (Z_0^2 + Z_{0,\text{ref}}^2)\sinh(\gamma L)}, \qquad (3.13)$$

where

$$\gamma = \alpha + \mathrm{j}\beta \text{ with } \beta = \frac{\omega}{v_\mathrm{p}}. \qquad (3.14)$$

3.7   TRANSMISSION LINE COMPONENTS: TWO-PORT SERIAL TRANSMISSION LINE STUB

Figure 3.7 shows the display of the Transmission Line Components: Two-Port Serial Transmission Line Stub device builder. The stub is connected to both ports of the Software VNA with zero transmission line lengths. The transmission line parameters including characteristic impedance, attenuation constant at 1 GHz, frequency dependence of the attenuation constant, phase velocity w.r.t. the speed of light in air and length can be inputted in the same way as that in Section 3.6. The termination load can be chosen to be a short circuit, open circuit or 50 Ω load.

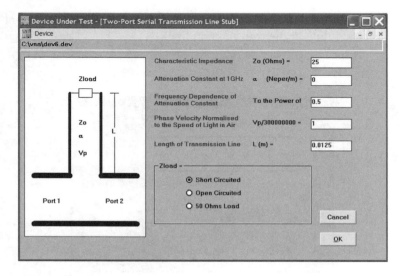

Figure 3.7   Transmission Line Components: Two-Port Serial Transmission Line Stub

The S-parameters of the serial transmission line stub with load impedance Z_L are given by

$$[S] = \begin{bmatrix} \dfrac{Z_{in}}{Z_{in} + 2Z_{0,ref}} & \dfrac{2Z_{0,ref}}{Z_{in} + 2Z_{0,ref}} \\ \dfrac{2Z_{0,ref}}{Z_{in} + 2Z_{0,ref}} & \dfrac{Z_{in}}{Z_{in} + 2Z_{0,ref}} \end{bmatrix}, \tag{3.15}$$

where

$$Z_{in} = Z_0 \frac{Z_L + Z_0 \tanh(\gamma L)}{Z_0 + Z_L \tanh(\gamma L)} \tag{3.16}$$

with $\gamma = \alpha + j\beta$ and $\beta = \omega/v_p$. For a short-circuit termination, $Z_L = 0\,\Omega$, and for an open-circuit termination, $Z_L = \infty\,\Omega$. The S-parameters of the serial transmission line stub can also be given by

$$[S] = \begin{bmatrix} \dfrac{1 + S_{11z}}{3 - S_{11z}} & \dfrac{2(1 - S_{11z})}{3 - S_{11z}} \\ \dfrac{2(1 - S_{11z})}{3 - S_{11z}} & \dfrac{1 + S_{11z}}{3 - S_{11z}} \end{bmatrix}, \tag{3.17}$$

where

$$S_{11z} = S_{11}^{(TL)} + \frac{S_{12}^{(TL)} S_{21}^{(TL)} \Gamma_L}{1 - S_{22}^{(TL)} \Gamma_L} \tag{3.18}$$

with $[S]^{(TL)}$ being the S-parameters of the transmission line section, i.e.

$$[S]^{(TL)} = \frac{\begin{bmatrix} (Z_0^2 - Z_{0,ref}^2)\sinh(\gamma L) & 2Z_0 Z_{0,ref} \\ 2Z_0 Z_{0,ref} & (Z_0^2 - Z_{0,ref}^2)\sinh(\gamma L) \end{bmatrix}}{2Z_0 Z_{0,ref} \cosh(\gamma L) + (Z_0^2 + Z_{0,ref}^2)\sinh(\gamma L)} \tag{3.19}$$

and

$$\Gamma_L = \frac{Z_L - Z_{0,ref}}{Z_L + Z_{0,ref}}. \tag{3.20}$$

3.8 TRANSMISSION LINE COMPONENTS: TWO-PORT PARALLEL TRANSMISSION LINE STUB

Figure 3.8 shows the display of the Transmission Line Components: Two-Port Parallel Transmission Line Stub device builder. The stub is connected to both ports of the Software VNA with zero transmission

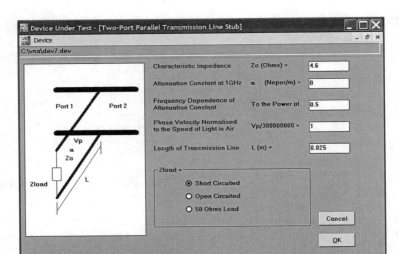

Figure 3.8   Transmission Line Components: Two-Port Parallel Transmission Line Stub

line lengths. The transmission line parameters including characteristic impedance, attenuation constant at 1 GHz, frequency dependence of the attenuation constant, phase velocity w.r.t. the speed of light in air and length can be inputted in the same way as that in Section 3.6. The termination load can again be chosen to be a short circuit, open circuit or 50 Ω load.

The S-parameters of the parallel transmission line stub with load impedance Z_L are given by

$$[S] = \begin{bmatrix} \dfrac{-Y_{\mathrm{in}} Z_{0,\mathrm{ref}}}{2 + Y_{\mathrm{in}} Z_{0,\mathrm{ref}}} & \dfrac{2}{2 + Y_{\mathrm{in}} Z_{0,\mathrm{ref}}} \\[3mm] \dfrac{2}{2 + Y_{\mathrm{in}} Z_{0,\mathrm{ref}}} & \dfrac{-Y_{\mathrm{in}} Z_{0,\mathrm{ref}}}{2 + Y_{\mathrm{in}} Z_{0,\mathrm{ref}}} \end{bmatrix}, \tag{3.21}$$

where

$$Y_{\mathrm{in}} = \left[Z_0 \frac{Z_L + Z_0 \tanh(\gamma L)}{Z_0 + Z_L \tanh(\gamma L)} \right]^{-1} \tag{3.22}$$

with $\gamma = \alpha + \mathrm{j}\beta$ and $\beta = \omega / v_{\mathrm{p}}$.

The S-parameters of the parallel transmission line stub can also be given by

$$[S] = \begin{bmatrix} \dfrac{S_{11z} - 1}{3 + S_{11z}} & \dfrac{2(1 + S_{11z})}{3 + S_{11z}} \\[3mm] \dfrac{2(1 + S_{11z})}{3 + S_{11z}} & \dfrac{S_{11z} - 1}{3 + S_{11z}} \end{bmatrix}, \tag{3.23}$$

where

$$S_{11z} = S_{11}^{(\mathrm{TL})} + \frac{S_{12}^{(\mathrm{TL})} S_{21}^{(\mathrm{TL})} \Gamma_{\mathrm{L}}}{1 - S_{22}^{(\mathrm{TL})} \Gamma_{\mathrm{L}}} \tag{3.24}$$

with $[S]^{(\mathrm{TL})}$ being the S-parameters of the transmission line section, i.e.

$$[S]^{(\mathrm{TL})} = \frac{\begin{bmatrix} (Z_0^2 - Z_{0,\mathrm{ref}}^2)\sinh(\gamma L) & 2Z_0 Z_{0,\mathrm{ref}} \\[2mm] 2Z_0 Z_{0,\mathrm{ref}} & (Z_0^2 - Z_{0,\mathrm{ref}}^2)\sinh(\gamma L) \end{bmatrix}}{2Z_0 Z_{0,\mathrm{ref}} \cosh(\gamma L) + (Z_0^2 + Z_{0,\mathrm{ref}}^2)\sinh(\gamma L)} \tag{3.25}$$

and

$$\Gamma_{\mathrm{L}} = \frac{Z_{\mathrm{L}} - Z_{0,\mathrm{ref}}}{Z_{\mathrm{L}} + Z_{0,\mathrm{ref}}}. \tag{3.26}$$

3.9 IDEAL TWO-PORT COMPONENTS: ATTENUATOR/GAIN BLOCK

Figure 3.9 shows the display of the Ideal Two-Port Components: Attenuator/Gain Block device builder. The device is connected to both ports of the Software VNA with zero transmission line lengths. The value of

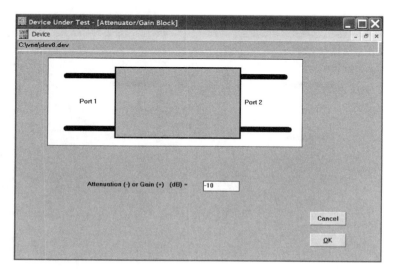

Figure 3.9 Ideal Two-Port Components: Attenuator/Gain Block

attenuation or gain in dB can be inputted. The S-parameters of the block are

$$[S] = \begin{bmatrix} 0 & e^{GdB/8.686} \\ e^{GdB/8.686} & 0 \end{bmatrix}, \tag{3.27}$$

where GdB is the gain in dB. If GdB is negative, the gain block becomes an attenuation block.

3.10 IDEAL TWO-PORT COMPONENTS: 1:N AND N:1 TRANSFORMER

Figure 3.10 shows the display of the Ideal Two-Port Components: 1:N and N:1 Transformer device builder. The device is connected to both ports of the Software VNA with zero transmission line lengths. A selection of 1:N or N:1 transformation can be made, and the value of N which can be greater or less than 1 can be inputted. The S-parameters of the 1:N transformer are

$$[S] = \begin{bmatrix} \dfrac{1 - N^2}{1 + N^2} & \dfrac{2N}{1 + N^2} \\[2ex] \dfrac{2N}{1 + N^2} & \dfrac{N^2 - 1}{1 + N^2} \end{bmatrix} \tag{3.28}$$

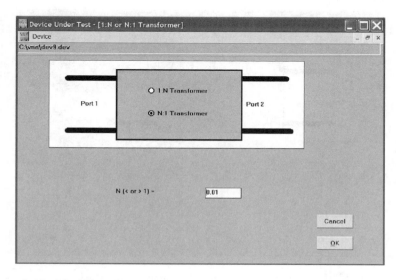

Figure 3.10 Ideal Two-Port Components: 1:N and N:1 Transformer

and those of the N:1 transformer are

$$[S] = \begin{bmatrix} \dfrac{N^2-1}{1+N^2} & \dfrac{2N}{1+N^2} \\[3mm] \dfrac{2N}{1+N^2} & \dfrac{1-N^2}{1+N^2} \end{bmatrix}. \tag{3.29}$$

3.11   IDEAL TWO-PORT COMPONENTS: ISOLATOR

Figure 3.11 shows the display of the Ideal Two-Port Components: Isolator device builder. The device is connected to both ports of the Software VNA with zero transmission line lengths. The S-parameters of the idealised isolator are

$$[S] = \begin{bmatrix} 0 & 0 \\ 1 & 0 \end{bmatrix}. \tag{3.30}$$

3.12   IDEAL TWO-PORT COMPONENTS: GYRATOR

Figure 3.12 shows the display of the Ideal Two-Port Components: Gyrator

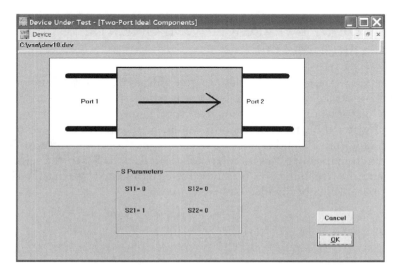

Figure 3.11   Ideal Two-Port Components: Isolator

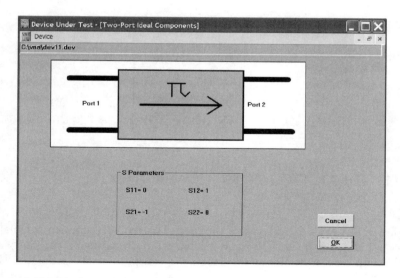

Figure 3.12   Ideal Two-Port Components: Gyrator

device builder. The device is connected to both ports of the Software VNA with zero transmission line lengths. The S-parameters of the idealised gyrator are

$$[S] = \begin{bmatrix} 0 & 1 \\ -1 & 0 \end{bmatrix}. \tag{3.31}$$

3.13   IDEAL TWO-PORT COMPONENTS: CIRCULATOR

Figure 3.13 shows the display of the Ideal Two-Port Components: Circulator device builder. The circulator is a three-port device. Any port of the circulator can be selected from the list to connect to Port 1 or Port 2 of the Software VNA. However, a port cannot be selected to connect to both Port 1 and Port 2 simultaneously. When this happens, the Software VNA would automatically select an alternative connection. When the selections are made, the remaining port is assumed to be terminated by a matched load. The S-parameters of the idealised circulator are

$$[S] = \begin{bmatrix} 0 & 0 & 1 \\ 1 & 0 & 0 \\ 0 & 1 & 0 \end{bmatrix}. \tag{3.32}$$

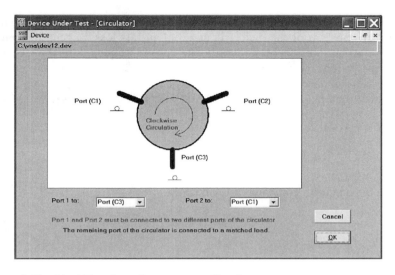

Figure 3.13   Ideal Two-Port Components: Circulator

3.14   PHYSICAL TRANSMISSION LINES: COAXIAL LINE

Figure 3.14 shows the display of the Physical Transmission Lines: Coaxial Line device builder. The coaxial line is a two-port device. The electrical parameters including the radii of inner and outer conductors, conductivity

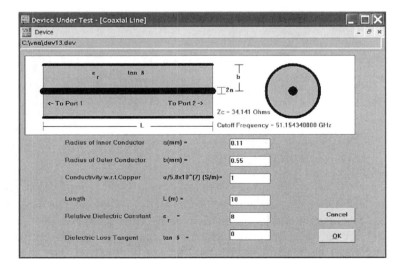

Figure 3.14   Physical Transmission Lines: Coaxial Line

of the conductor with respect to that of copper, i.e. 5.8×10^8 S/m, relative dielectric constant of the dielectric material and its loss tangent and length of the coaxial line can be inputted. The cut-off frequency of the structure for TEM propagation will be calculated when either one of the dimensions or the relative dielectric constant is changed.

The characteristic impedance of the coaxial line is given by (Pozar, 1990; Wadell, 1991)

$$Z_0 = \frac{60}{\sqrt{\varepsilon_r}} \ln\left(\frac{b}{a}\right). \tag{3.33}$$

The propagation constant of the coaxial line is

$$\gamma = \alpha + j\beta, \tag{3.34}$$

where

$$\alpha = \frac{R_s}{4\pi} \frac{1}{Z_0}\left(\frac{1}{a} + \frac{1}{b}\right) + \frac{\omega}{2c}\sqrt{\varepsilon_r}\tan\delta \quad \text{and} \quad \beta = \frac{\omega}{c}\sqrt{\varepsilon_r} \tag{3.35}$$

with $R_s = \sqrt{\omega\mu_0/2\sigma}$ and c is the speed of light in air. The cut-off frequency for TEM propagation is calculated by (Pozar, 1990; Wadell, 1991)

$$f_{\text{cutoff}} = \frac{c}{\pi(a+b)\sqrt{\varepsilon_r}}. \tag{3.36}$$

The S-parameters of the coaxial line can be obtained using the above parameters and the equations for the general transmission line described in Section 3.6, i.e.

$$[S] = \frac{\begin{bmatrix} (Z_0^2 - Z_{0,\text{ref}}^2)\sinh(\gamma L) & 2Z_0 Z_{0,\text{ref}} \\ 2Z_0 Z_{0,\text{ref}} & (Z_0^2 - Z_{0,\text{ref}}^2)\sinh(\gamma L) \end{bmatrix}}{2Z_0 Z_{0,\text{ref}}\cosh(\gamma L) + (Z_0^2 + Z_{0,\text{ref}}^2)\sinh(\gamma L)}. \tag{3.37}$$

3.15 PHYSICAL TRANSMISSION LINES: MICROSTRIP LINE

Figure 3.15 shows the display of the Physical Transmission Lines: Microstrip Line device builder. The microstrip line is a two-port device. The width and thickness of the strip and its conductivity w.r.t. that of copper, thickness of the dielectric substrate, relative dielectric constant and loss tangent of the substrate and length of the microstrip line can be inputted. The radiation loss of the microstrip line may be included in the modelling. The cut-off frequency of the structure for quasi-TEM propagation will be calculated

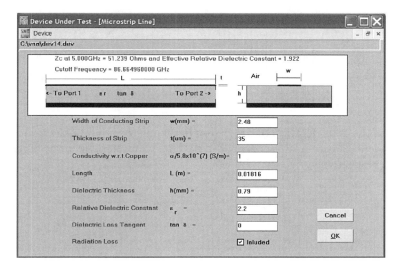

Figure 3.15   Physical Transmission Lines: Microstrip Line

when either one of the dimensions or the relative dielectric constant is changed.

The characteristic impedance of the microstrip line is given by the following equations (Bahl, 1989):

$$
Z_0 = \begin{cases}
\dfrac{60}{\sqrt{\varepsilon_{\mathrm{re}}}} \ln\left(\dfrac{8h}{w'} + 0.25\dfrac{w'}{h}\right) & \text{for } w/h < 1 \\[3mm]
\dfrac{120\pi}{\sqrt{\varepsilon_{\mathrm{re}}}} \left[\dfrac{w'}{h} + 1.393 + 0.667\ln\left(\dfrac{w'}{h} + 1.444\right)\right]^{-1} & \text{for } w/h \geq 1
\end{cases},
\tag{3.38}
$$

where

$$
\frac{w'}{h} = \begin{cases}
\dfrac{w}{h} + \dfrac{1.25}{\pi}\dfrac{t}{h}\left(1 + \ln\left(\dfrac{4\pi w}{t}\right)\right) & \text{for } \dfrac{w}{h} < \dfrac{1}{2\pi} \\[3mm]
\dfrac{w}{h} + \dfrac{1.25}{\pi}\dfrac{t}{h}\left(1 + \ln\left(\dfrac{2h}{t}\right)\right) & \text{for } \dfrac{w}{h} \geq \dfrac{1}{2\pi}
\end{cases}
\tag{3.39}
$$

and the effective dielectric constant

$$
\varepsilon_{\mathrm{re}} = \frac{\varepsilon_{\mathrm{r}} + 1}{2} + \frac{\varepsilon_{\mathrm{r}} - 1}{2} F(w/h) - \frac{\varepsilon_{\mathrm{r}} - 1}{4.6}\frac{t}{h}\left(\frac{w}{h}\right)^{-1/2}
\tag{3.40}
$$

with

$$F(w/h) = \begin{cases} \left(1 + \dfrac{12h}{w}\right)^{-0.5} + 0.04\left(1 - \dfrac{w}{h}\right)^2 & \text{for } \dfrac{w}{h} < 1 \\ \left(1 + \dfrac{12h}{w}\right)^{-0.5} & \text{for } \dfrac{w}{h} > 1 \end{cases}. \tag{3.41}$$

With the consideration of dispersion in the microstrip line, the frequency-dependent effective dielectric constant and characteristic impedance are given, respectively, by

$$\varepsilon_{re}(f) = \left(\frac{\sqrt{\varepsilon_r} - \sqrt{\varepsilon_{re}}}{1 + 4F^{-1.5}} + \sqrt{\varepsilon_{re}}\right) \tag{3.42}$$

and

$$Z_0(f) = Z_0 \frac{\varepsilon_{re}(f) - 1}{\varepsilon_{re} - 1} \sqrt{\frac{\varepsilon_{re}}{\varepsilon_{re}(f)}}, \tag{3.43}$$

where

$$F = \frac{4hf\sqrt{\varepsilon_r - 1}}{c} \left\{0.5 + \left[1 + 2\ln\left(1 + \frac{w}{h}\right)\right]^2\right\} \tag{3.44}$$

with $c = 3 \times 10^8$ m/s.

The attenuation constant α includes conducting, dielectric and radiation losses, i.e.

$$\alpha = \alpha_c + \alpha_d + \alpha_r \tag{3.45}$$

with a unit of Neper/m. The attenuation constant α_c is given by

$$\alpha_c = \begin{cases} 0.1588 \dfrac{R_s}{hZ_0} \dfrac{32 - (w'/h)^2}{32 + (w'/h)^2} A & \text{for } w/h < 1 \\ 0.70228 \times 10^{-5} \dfrac{R_s Z_0 \varepsilon_{re}}{h} \left[\dfrac{w'}{h} + \dfrac{0.667w'/h}{w/h + 1.444}\right] A & \text{for } w/h \geq 1 \end{cases} \tag{3.46}$$

with

$$A = \begin{cases} 1 + \dfrac{h}{w'}\left[1 + \dfrac{1.25t}{\pi w} + \dfrac{1.25}{\pi}\ln\left(\dfrac{4\pi w}{t}\right)\right] & \text{for } \dfrac{w}{h} < \dfrac{1}{2\pi} \\ 1 + \dfrac{h}{w'}\left[1 - \dfrac{1.25t}{\pi h} + \dfrac{1.25}{\pi}\ln\left(\dfrac{2h}{t}\right)\right] & \text{for } \dfrac{w}{h} \geq \dfrac{1}{2\pi} \end{cases} \tag{3.47}$$

and

$$R_s = \sqrt{\frac{\omega\mu_0}{2\sigma}}. \tag{3.48}$$

The attenuation constant α_d is given by

$$\alpha_d = \frac{\pi f}{c} \frac{\varepsilon_r}{\sqrt{\varepsilon_{re}(f)}} \frac{\varepsilon_{re}(f) - 1}{\varepsilon_r - 1} \tan\delta \tag{3.49}$$

and α_r by

$$\alpha_r = \frac{\pi f}{cQ_r} = \frac{\pi f}{c}\left(\frac{480\pi(hf/c)^2 R\sqrt{\varepsilon_{re}(f)}}{Z_0}\right) \tag{3.50}$$

with

$$R = \frac{\varepsilon_{re}(f) + 1}{\varepsilon_{re}(f)} - \frac{(\varepsilon_{re}(f) - 1)^2}{2\varepsilon_{re}(f)^{3/2}} \ln\frac{\sqrt{\varepsilon_{re}(f)} + 1}{\sqrt{\varepsilon_{re}(f)} - 1}. \tag{3.51}$$

The cut-off frequency for quasi-TEM propagation is estimated by (Wadell, 1991)

$$f_{\text{cutoff}} = \begin{cases} \dfrac{c}{4h\sqrt{\varepsilon_r - 1}} & \text{for } \varepsilon_r > 2 \\ \dfrac{c}{4h} & \text{for } \varepsilon_r \le 2 \end{cases}. \tag{3.52}$$

Using the above equations for the characteristic impedance and the propagation constant

$$\gamma = \alpha + j\beta \text{ with } \beta = \frac{2\pi f}{c}\sqrt{\varepsilon_{re}(f)}, \tag{3.53}$$

the S-parameters of the microstrip line can be obtained using the equations for the general transmission line described in Section 3.6, i.e.

$$[S] = \frac{\begin{bmatrix} \left(Z_0^2 - Z_{0,\text{ref}}^2\right)\sinh(\gamma L) & 2Z_0 Z_{0,\text{ref}} \\ 2Z_0 Z_{0,\text{ref}} & \left(Z_0^2 - Z_{0,\text{ref}}^2\right)\sinh(\gamma L) \end{bmatrix}}{2Z_0 Z_{0,\text{ref}}\cosh(\gamma L) + \left(Z_0^2 + Z_{0,\text{ref}}^2\right)\sinh(\gamma L)}. \tag{3.54}$$

3.16 PHYSICAL TRANSMISSION LINES: STRIPLINE

Figure 3.16 shows the display of the Physical Transmission Lines: Stripline device builder. The stripline is a two-port device. The width and thickness of the strip and its conductivity w.r.t. that of copper, thickness of dielectric substrate, relative dielectric constant and loss tangent of the substrate and length of the stripline can be inputted. The cut-off frequency of the structure for TEM propagation will be calculated when either one of the dimensions or the relative dielectric constant is changed.

The characteristic impedance of the stripline is given by the following equations (Pozar, 1990; Wadell, 1991; Bahl, 1989; Howe, 1974):

$$Z_0 = \frac{30\pi}{\sqrt{\varepsilon_r}} \frac{b}{w_e + 0.441b}, \tag{3.55}$$

where

$$\frac{w_e}{b} = \begin{cases} \dfrac{w}{b} & \text{for } \dfrac{w}{b} > 0.35 \\ \dfrac{w}{b} - \left(0.35 - \dfrac{w}{b}\right)^2 & \text{for } \dfrac{w}{b} \le 0.35 \end{cases}. \tag{3.56}$$

The attenuation constant α includes both conducting and dielectric losses, i.e.

$$\alpha = \alpha_c + \alpha_d. \tag{3.57}$$

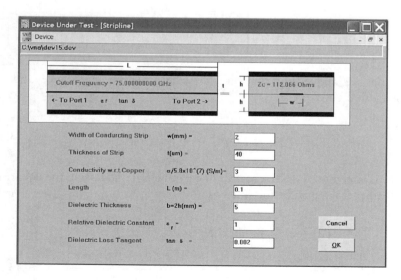

Figure 3.16 Physical Transmission Lines: Stripline

The attenuation constant α_c is given by

$$\alpha_c = \begin{cases} \dfrac{2.7 \times 10^{-3} R_s \varepsilon_r Z_0}{30\pi(b-t)} A & \text{for } \sqrt{\varepsilon_r} Z_0 < 120 \\ \dfrac{0.16 R_s}{Z_0 b} B & \text{for } \sqrt{\varepsilon_r} Z_0 \geq 120 \end{cases}, \tag{3.58}$$

where

$$\begin{aligned} A &= 1 + \frac{2w}{b-t} + \frac{1}{\pi} \frac{b+t}{b-t} \ln\left(\frac{2b-t}{t}\right), \\ B &= 1 + \frac{b}{(0.5w + 0.7t)} \left(0.5 + \frac{0.414t}{w} + \frac{1}{2\pi} \ln\left(\frac{4\pi w}{t}\right)\right) \end{aligned} \tag{3.59}$$

and

$$R_s = \sqrt{\frac{\omega \mu_0}{2\sigma}}. \tag{3.60}$$

The attenuation constant α_d is given by

$$\alpha_d = \frac{\pi f}{c} \sqrt{\varepsilon_r} \tan \delta, \tag{3.61}$$

where c is the speed of light in air. The cut-off frequency for TEM propagation is estimated by (Wadell, 1991)

$$f_{\text{cutoff}} = \frac{c}{2w\sqrt{\varepsilon_r}}. \tag{3.62}$$

Using the above equations for the characteristic impedance and the propagation constant given by

$$\gamma = \alpha + j\beta \text{ with } \beta = \frac{2\pi f}{c} \sqrt{\varepsilon_r}, \tag{3.63}$$

the S-parameters of the stripline can be obtained using the equations for the general transmission line described in Section 3.6, i.e.

$$[S] = \frac{\begin{bmatrix} (Z_0^2 - Z_{0,\text{ref}}^2) \sinh(\gamma L) & 2Z_0 Z_{0,\text{ref}} \\ 2Z_0 Z_{0,\text{ref}} & (Z_0^2 - Z_{0,\text{ref}}^2) \sinh(\gamma L) \end{bmatrix}}{2Z_0 Z_{0,\text{ref}} \cosh(\gamma L) + (Z_0^2 + Z_{0,\text{ref}}^2) \sinh(\gamma L)}. \tag{3.64}$$

3.17 PHYSICAL TRANSMISSION LINES: COPLANAR WAVEGUIDE

Figure 3.17 shows the display of the Physical Transmission Lines: Coplanar Waveguide device builder. The coplanar waveguide is a two-port device. The width and thickness of the strip and its conductivity w.r.t. that of copper, gap between strips, thickness of dielectric substrate, relative dielectric constant and loss tangent of the substrate and length of the coplanar waveguide can be inputted. The radiation loss of the structure may be included in the modelling. The cut-off frequency of the structure for quasi-TEM propagation will be calculated when either one of the dimensions or the relative dielectric constant is changed.

The characteristic impedance of the coplanar waveguide is given by (Gupta, Garg, Bahl and Bhartia, 1996)

$$Z_0 = \frac{30\pi}{\sqrt{\varepsilon_{re}}} \frac{K'(k_e)}{K(k_e)}, \tag{3.65}$$

where $K(k_e)$ and $K'(k_e)$ are the complete elliptic integrals of the first kind and its complement, respectively, with argument k_e given by

$$k_e = \frac{s + (1.25t/\pi)[1 + \ln(4\pi s/t)]}{s + 2w - (1.25t/\pi)[1 + \ln(4\pi s/t)]} \tag{3.66}$$

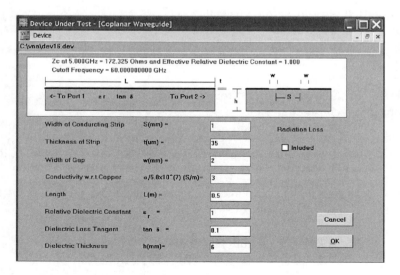

Figure 3.17 Physical Transmission Lines: Coplanar Waveguide

and $K'(k_e) = K(\sqrt{1-k_e^2})$. The effective relative dielectric constant ε_{re} is given by

$$\varepsilon_{re} = 1 + \frac{\varepsilon_r - 1}{2} \frac{K(k_2)}{K'(k_2)} \frac{K'(k_1)}{K(k_1)} \tag{3.67}$$

with

$$k_1 = \frac{s}{s+2w} \quad \text{and} \quad k_2 = \frac{\sinh\left(\dfrac{\pi s}{4h}\right)}{\sinh\left(\dfrac{\pi(s+2w)}{4h}\right)}. \tag{3.68}$$

The elliptic functions can be approximated by (Gupta, Garg, Bahl and Bhartia, 1996)

$$K(k) = \begin{cases} \dfrac{\pi}{2}\{1 + 2(k/8) + 9(k/8)^2 \\ \quad + 50(k/8)^3 + 306.25(k/8)^4 + \cdots\} & \text{for } 0 \le k < 0.07 \\[6pt] p + (p-1)(k'2/4) + 9(p - 7/6)(k'4/64) \\ \quad + 25(p - 137/30)(k'6/256) + \cdots & \text{for } 0.707 \le k \le 1 \end{cases} \tag{3.69}$$

with

$$k' = \sqrt{1 - k^2} \quad \text{and} \quad p = \ln(4/k') \tag{3.70}$$

or

$$\frac{K(k)}{K'(k)} = \begin{cases} \dfrac{\pi}{\ln[2(1+\sqrt{k'})/(1-\sqrt{k'})]} & \text{for } 0 \le k < 0.07 \\[6pt] \dfrac{\ln[2(1+\sqrt{k})/(1-\sqrt{k})]}{\pi} & \text{for } 0.707 \le k \le 1 \end{cases}. \tag{3.71}$$

The attenuation constant α includes conducting, dielectric and radiation losses, i.e.

$$\alpha = \alpha_c + \alpha_d + \alpha_r. \tag{3.72}$$

The attenuation constant α_c is given by

$$\alpha_c = \frac{R_s\sqrt{\varepsilon_{re}}}{480\pi K'(k_1)(1-k_1^2)} \left\{ \frac{2}{s}\left[\pi + \ln\left(\frac{4\pi s(1-k_1^2)}{t(1+k_1^2)}\right)\right] \right. \\ \left. + \frac{2}{s+2w}\left[\pi + \ln\left(\frac{4\pi(s+2w)(1-k_1^2)}{t(1+k_1^2)}\right)\right]\right\} \tag{3.73}$$

with

$$R_s = \sqrt{\frac{\omega \mu_0}{2\sigma}}. \tag{3.74}$$

The attenuation constant α_d is given by

$$\alpha_d = \frac{\pi f}{c} \frac{\varepsilon_r}{\sqrt{\varepsilon_{re}}} \frac{\varepsilon_{re} - 1}{\varepsilon_r - 1} \tan \delta, \tag{3.75}$$

where c is the speed of light in air, and the attenuation constant α_r is given by

$$\alpha_r = 2 \left(\frac{\pi}{2}\right)^5 \frac{(1 - \varepsilon_{re}/\varepsilon_r)^2}{\sqrt{\varepsilon_{re}/\varepsilon_r}} \frac{(s + 2w)^2}{K(k_1)K'(k_1)} \left(\frac{f\sqrt{\varepsilon_r}}{c}\right)^3. \tag{3.76}$$

The cut-off frequency for quasi-TEM propagation is estimated by (Wadell, 1991)

$$f_{cutoff} = \frac{c}{(s + 2w)\sqrt{(\varepsilon_r + 1)/2}}. \tag{3.77}$$

Using the above equations for the characteristic impedance and the propagation constant

$$\gamma = \alpha + j\beta \text{ with } \beta = \frac{2\pi f}{c} \sqrt{\varepsilon_{re}}, \tag{3.78}$$

the S-parameters of the coplanar waveguide can be obtained using the equations for the general transmission line described in Section 3.6, i.e.

$$[S] = \frac{\begin{bmatrix} (Z_0^2 - Z_{0,ref}^2) \sinh(\gamma L) & 2Z_0 Z_{0,ref} \\ 2Z_0 Z_{0,ref} & (Z_0^2 - Z_{0,ref}^2) \sinh(\gamma L) \end{bmatrix}}{2Z_0 Z_{0,ref} \cosh(\gamma L) + (Z_0^2 + Z_{0,ref}^2) \sinh(\gamma L)}. \tag{3.79}$$

3.18   PHYSICAL TRANSMISSION LINES: COPLANAR STRIPS

Figure 3.18 shows the display of the Physical Transmission Lines: Coplanar Strips device builder. The coplanar strip is a two-port device. The width and thickness of the strips and their conductivity w.r.t. that of copper, gap between strips, thickness of dielectric substrate, relative dielectric constant and loss tangent of the substrate and length of the coplanar strips can be inputted. The radiation loss of the structure may be included in the modelling. The cut-off frequency of the structure for quasi-TEM

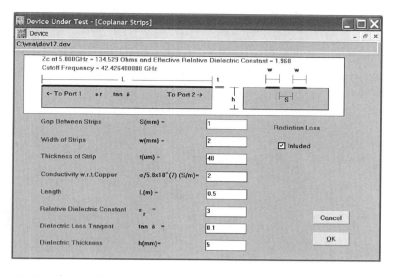

Figure 3.18   Physical Transmission Lines: Coplanar Strips

propagation will be calculated when either one of the dimensions or the relative dielectric constant is changed.

The characteristic impedance of the coplanar strips is given by (Gupta, Garg, Bahl and Bhartia, 1996)

$$Z_0 = \frac{120\pi}{\sqrt{\varepsilon_{re}}} \frac{K(k_1)}{K'(k_1)},$$ (3.80)

where $K(k_1)$ and $K'(k_1)$ have been given in Section 3.17. The argument k_1 is defined as

$$k_1 = \frac{s}{s+2w}.$$ (3.81)

The effective relative dielectric constant ε_{re} is given by

$$\varepsilon_{re} = 1 + \frac{\varepsilon_r - 1}{2} \frac{K(k_2)}{K'(k_2)} \frac{K'(k_1)}{K(k_1)}$$ (3.82)

with

$$k_2 = \frac{\sinh\left(\dfrac{\pi s}{4h}\right)}{\sinh\left(\dfrac{\pi(s+2w)}{4h}\right)}.$$ (3.83)

The attenuation constant α includes conducting, dielectric and radiation losses, i.e.

$$\alpha = \alpha_c + \alpha_d + \alpha_r.$$ (3.84)

The attenuation constant α_c is given by,

$$\alpha_c = \frac{R_s\sqrt{\varepsilon_{re}}}{480\pi K'(k_1)(1-k_1^2)} \left\{ \frac{2}{s}\left[\pi + \ln\left(\frac{4\pi s(1-k_1^2)}{t(1+k_1^2)}\right)\right] \right.$$
$$\left. + \frac{2}{s+2w}\left[\pi + \ln\left(\frac{4\pi(s+2w)(1-k_1^2)}{t(1+k_1^2)}\right)\right]\right\}$$

(3.85)

with

$$R_s = \sqrt{\frac{\omega\mu_0}{2\sigma}}.$$

(3.86)

The attenuation constant α_d is given by

$$\alpha_d = \frac{\pi f}{c}\frac{\varepsilon_r}{\sqrt{\varepsilon_{re}}}\frac{\varepsilon_{re}-1}{\varepsilon_r-1}\tan\delta,$$

(3.87)

where c is the speed of light in air, and the attenuation constant α_r is given by

$$\alpha_r = (3-\sqrt{8})\left(\frac{\pi}{2}\right)^2\sqrt{\frac{\varepsilon_{re}}{\varepsilon_r}}\left(\frac{\varepsilon_r-1}{2\varepsilon_r}\right)^2\frac{w^2}{K(k_1)K'(k_1)}\left(\frac{f\sqrt{\varepsilon_r}}{c}\right)^3.$$

(3.88)

The cut-off frequency for quasi-TEM propagation is estimated by (Wadell, 1991)

$$f_{cutoff} = \frac{c}{(s+2w)\sqrt{(\varepsilon_r+1)/2}}.$$

(3.89)

Using the above equations for the characteristic impedance and the propagation constant given by

$$\gamma = \alpha + j\beta \text{ with } \beta = \frac{2\pi f}{c}\sqrt{\varepsilon_{re}},$$

(3.90)

the S-parameters of the coplanar strip can be obtained using the equations for the general transmission line described in Section 3.6, i.e.

$$[S] = \frac{\begin{bmatrix} (Z_0^2 - Z_{0,ref}^2)\sinh(\gamma L) & 2Z_0 Z_{0,ref} \\ 2Z_0 Z_{0,ref} & (Z_0^2 - Z_{0,ref}^2)\sinh(\gamma L) \end{bmatrix}}{2Z_0 Z_{0,ref}\cosh(\gamma L) + (Z_0^2 + Z_{0,ref}^2)\sinh(\gamma L)}.$$

(3.91)

3.19   PHYSICAL LINE DISCONTINUITIES: COAXIAL LINE DISCONTINUITIES

Figure 3.19(a)–(c) shows the coaxial step, gap and open-end discontinuities, respectively, together with their equivalent circuits. The models of these discontinuities are described below.

3.19.1   Step Discontinuity

The step discontinuity can be represented by a shunt capacitance C_{eq}. The shunt capacitance is modelled by (Wadell, 1991)

$$C_{eq} = (\varepsilon_{r1} + \varepsilon_{r2})\, \pi r_1 C_{d2} + (\varepsilon_{r1} + \varepsilon_{r2})\, \pi r_{1av} C_{d1}, \tag{3.92}$$

where C_{d1} is the capacitance for the inner step and C_{d2} the capacitance for the outer step. They are given by

$$
\begin{aligned}
C_{d1} &= \frac{\varepsilon_0}{\pi}\left[\frac{\alpha_1^2+1}{\alpha_1}\ln\frac{1+\alpha_1}{1-\alpha_1} - 2\ln\frac{4\alpha_1}{1-\alpha_1^2}\right] + 1.11\times 10^{-13}(1-\alpha_1)(\tau_1 - 1)\ \text{(F/m)},\\
C_{d2} &= \frac{\varepsilon_0}{\pi}\left[\frac{\alpha_2^2+1}{\alpha_2}\ln\frac{1+\alpha_2}{1-\alpha_2} - 2\ln\frac{4\alpha_2}{1-\alpha_2^2}\right] + 4.12\times 10^{-13}(1-\alpha_2)(\tau_2 - 1.4)\ \text{(F/m)},
\end{aligned}
\tag{3.93}
$$

with

$$
\begin{aligned}
r_{1av} &= \frac{r_0 + r_1}{2}, \qquad \alpha_2 = \frac{r_2 - r_{1av}}{r_3 - r_{1av}} \quad\text{and}\quad \tau_2 = \frac{r_3}{r_{1av}}\\
r_{3av} &= \frac{r_3 + r_2}{2}, \qquad \alpha_1 = \frac{r_{3av} - r_1}{r_{3av} - r_0} \quad\text{and}\quad \tau_1 = \frac{r_{3av}}{r_0}.
\end{aligned}
\tag{3.94}
$$

It should be noted that the factor $(1 - \alpha_2)$ instead of $(0.8 - \alpha_2)$ in Equation (3.93) is used in C_{d2} so that the formulas are applicable to the range $0.01 \le \alpha \le 1$ and $1 \le \tau \le 6.0$ for both outer and inner steps.

The S-parameters of the step discontinuity can then be obtained using the equations for a two-port shunt admittance device described in Section 3.5, i.e.

$$[S] = \begin{bmatrix} \dfrac{-YZ_{0,ref}}{2 + YZ_{0,ref}} & \dfrac{2}{2 + YZ_{0,ref}} \\[2ex] \dfrac{2}{2 + YZ_{0,ref}} & \dfrac{-YZ_{0,ref}}{2 + YZ_{0,ref}} \end{bmatrix} \tag{3.95}$$

with

$$Y = j\omega C_{eq}. \tag{3.96}$$

3.19.2   Gap Discontinuity

The gap discontinuity can be modelled as a series capacitor. The equivalent series capacitance C_{eq} is given by

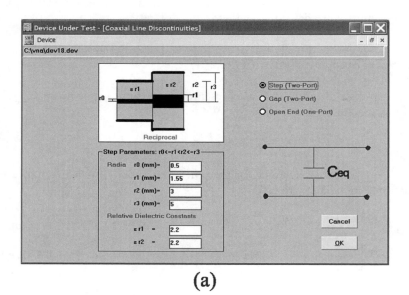

(a)

(b)

Figure 3.19   Physical Line Discontinuities: Coaxial Line Discontinuities: (a) step, (b) gap and (c) open end

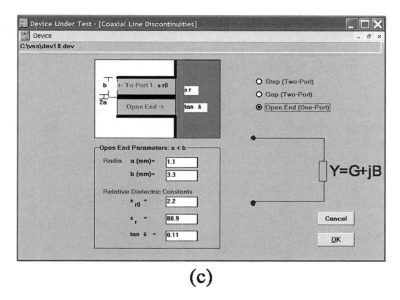

(c)

Figure 3.19 (Continued)

$$C_{eq} = \varepsilon_0 \varepsilon_r \frac{\pi a^2}{s}.$$ (3.97)

The S-parameters of the gap discontinuity can then be obtained using the equations for a two-port series impedance device described in Section 3.4, i.e.

$$[S] = \begin{bmatrix} \dfrac{Z}{Z + 2Z_{0,\text{ref}}} & \dfrac{2Z_{0,\text{ref}}}{Z + 2Z_{0,\text{ref}}} \\ \dfrac{2Z_{0,\text{ref}}}{Z + 2Z_{0,\text{ref}}} & \dfrac{Z}{Z + 2Z_{0,\text{ref}}} \end{bmatrix}$$ (3.98)

with

$$Z = \frac{1}{j\omega C_{eq}}.$$ (3.99)

3.19.3   Open-End Discontinuity

The open-end discontinuity can be modelled as an equivalent admittance (Misra et al., 1990; Marcuvitz, 1951), which is given by $Y = G + jB$ with

$$G = \frac{2}{3}\left(\frac{k_0\sqrt{\varepsilon_{\text{rout}}}}{2}\right)^4 (b^2 - a^2)^2 \frac{Y_0\sqrt{\varepsilon_{\text{rout}}}}{\ln(b/a)\sqrt{\varepsilon_{\text{rin}}}}$$ (3.100)

and

$$B = \frac{2Y_0 k_0 \varepsilon_{\text{rout}}}{\pi \ln(b/a)\sqrt{\varepsilon_{\text{rin}}}} \left[(a^2+b^2)^{1/2} \pi - 2(a+b) \right], \tag{3.101}$$

where $Y_0 = \sqrt{\varepsilon_0/\mu_0} = 1/120\pi$. These formulas are valid for $k_0\sqrt{\varepsilon_r}b \le 1$. If the external medium is of low loss with loss tangent $\tan\delta$, the G and B values can be approximated to

$$G' = G(1-j\tan\delta)^{5/2} = G\left(1-j\frac{5}{2}\tan\delta\right), \quad B' = B(1-j\tan\delta) \tag{3.102}$$

so that

$$Y = G' + jB' = (G+B\tan\delta) + j\left(B - \frac{5}{2}G\tan\delta\right). \tag{3.103}$$

The S-parameter of the open-end discontinuity can then be obtained using the equations for a one-port terminated load described in Section 3.3, i.e.

$$S_{11} = \frac{Z_L - Z_{0,\text{ref}}}{Z_L + Z_{0,\text{ref}}} \tag{3.104}$$

with

$$Z_L = 1/Y. \tag{3.105}$$

3.20   PHYSICAL LINE DISCONTINUITIES: MICROSTRIP LINE DISCONTINUITIES

Figure 3.20(a)–(e) shows, respectively, the microstrip step, gap, bend, slit and open-end discontinuities together with their equivalent circuits. The models of these discontinuities are described below.

3.20.1   Step Discontinuity

The step discontinuity can be represented by series inductance L_1, shunt capacitance C_s and series inductance L_2 (Gupta, Garg, Bahl and

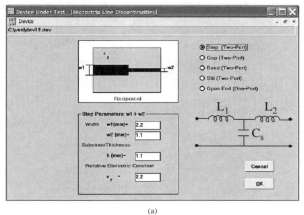

(a)

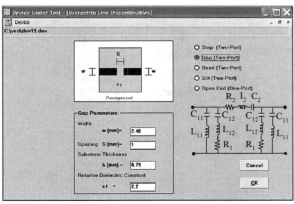

(b)

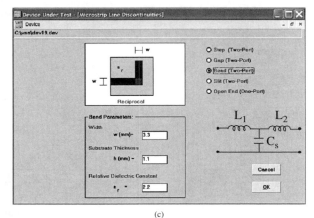

(c)

Figure 3.20   Physical Line Discontinuities: Microstrip Line Discontinuities: (a) step, (b) gap, (c) bend, (d) slit and (e) open end

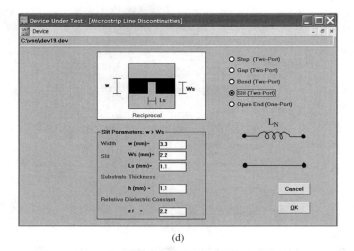

(d)

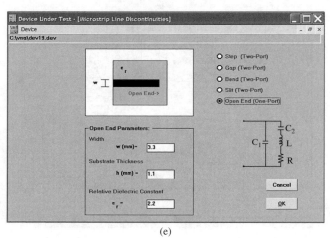

(e)

Figure 3.20 (Continued)

Bhartia, 1996), where

$$L_1 = \left(\frac{Z_{0m1}\sqrt{\varepsilon_{re1}}}{Z_{0m1}\sqrt{\varepsilon_{re1}} + Z_{0m2}\sqrt{\varepsilon_{re2}}} \right) L_s, \quad L_2 = \left(\frac{Z_{0m2}\sqrt{\varepsilon_{re2}}}{Z_{0m1}\sqrt{\varepsilon_{re1}} + Z_{0m2}\sqrt{\varepsilon_{re2}}} \right) L_s;$$

$$C_s = 1370h\frac{\sqrt{\varepsilon_{re1}}}{Z_{0m1}}\left(1 - \frac{w_2}{w_1}\right)\left(\frac{\varepsilon_{re1} + 0.3}{\varepsilon_{re1} - 0.258}\right)\left(\frac{w_1/h + 0.264}{w_1/h + 0.8}\right) \text{(pF)}$$

$$(3.106)$$

and

$$L_s = 987h \left(1 - \frac{Z_{0m1}}{Z_{0m2}} \sqrt{\frac{\varepsilon_{re1}}{\varepsilon_{re2}}}\right)^2 \text{ (nH)} \tag{3.107}$$

with Z_{0m} and ε_{re} values calculated using the microstrip line equations in Section 3.15 for conductor thickness $t = 0$.

The S-parameters of the step discontinuity can then be obtained using the equations for a 'T' impedance network in Table 1.8, Section 1.5, i.e.

$$[S] = \frac{\begin{bmatrix} -Z_{0,ref}^2 + (Z_1 - Z_2)Z_{0,ref} + Z_1Z_2 + Z_2Z_3 + Z_3Z_1 & 2Z_{0,ref}Z_3 \\ 2Z_{0,ref}Z_3 & -Z_{0,ref}^2 + (Z_2 - Z_1)Z_{0,ref} + Z_1Z_2 + Z_2Z_3 + Z_3Z_1 \end{bmatrix}}{Z_{0,ref}^2 + (Z_1 + Z_2 + 2Z_3)Z_{0,ref} + Z_1Z_2 + Z_2Z_3 + Z_3Z_1} \tag{3.108}$$

with

$$Z_1 = j\omega L_1, \quad Z_2 = j\omega L_2 \quad \text{and} \quad Z_3 = 1/(j\omega C_s). \tag{3.109}$$

3.20.2   Gap Discontinuity

The model used to describe the gap discontinuity is that by Alexopolous and Wu (1994). The R, L and C values of the equivalent circuit shown in Figure 3.20(b) are given by

$$C_{11} \times \frac{25Z_0}{h} = \left[1.125 \tanh\left(1.358\frac{w}{h}\right) - 0.315\right] \tanh\left[\left(0.0262 + 0.184\frac{h}{w}\right)\right.$$
$$\left. + \left(0.217 + 0.0619 \ln\frac{w}{h}\right)\frac{S}{h}\right] \text{ (pF }\Omega\text{)},$$

$$C_{12} \times \frac{25Z_0}{h} = \left[6.832 \tanh\left(0.0109\frac{w}{h}\right) + 0.910\right] \tanh\left[\left(1.411 + 0.314\frac{h}{w}\right)\right.$$
$$\left. + \left(\frac{S}{h}\right)^{1.248 + 0.360 \tan^{-1} w/h}\right] \text{ (pF }\Omega\text{)},$$

$$L_{11} \times \frac{25}{hZ_0} = \left[0.134 + 0.0436 \ln\frac{h}{w}\right] \exp\left[-1 \times \left(3.656 + 0.246\frac{h}{w}\right)\right.$$
$$\left. \left(\frac{S}{h}\right)^{1.739 + 0.390 \ln w/h}\right] \text{ (nH/}\Omega\text{)},$$

$$L_{12} \times \frac{25}{hZ_0} = \left[0.008285\tanh\left(0.5665\frac{w}{h}\right) + 0.0103\right]$$
$$+ \left[0.1827 + 0.00715\ln\frac{w}{h}\right]$$
$$\exp\left[-1 \times \left(5.207 + 1.283\tanh\left(1.656\frac{h}{w}\right)\right)\left(\frac{S}{h}\right)^{0.542 + 0.873\tan^{-1} w/h}\right]$$
$$(\text{nH}/\Omega),$$

$$R_1/Z_0 = 1.024\tanh\left(2.025\frac{w}{h}\right)\tanh\left[\left(0.01584 + 0.0187\frac{h}{w}\right)\frac{S}{h}\right.$$
$$\left. + \left(0.1246 + 0.0394\sinh\left(\frac{w}{h}\right)\right)\right],$$

$$C_2 \times \frac{25Z_0}{h} = \left[0.1776 + 0.05104\ln\left(\frac{w}{h}\right)\right]\frac{h}{S} + \left[0.574 + 0.3615\frac{h}{w}\right.$$
$$\left. + 1.156\ln\left(\frac{w}{h}\right)\right]\text{sech}\left(2.3345\frac{S}{h}\right)(\text{pF}\,\Omega),$$

$$L_2 \times \frac{25}{hZ_0} = \left[0.00228 + \frac{0.0873}{7.52w/h + \cosh(w/h)}\right]$$
$$\sinh\left(2.3345\frac{S}{h}\right)(\text{nH}/\Omega),\ (\text{nH}/\Omega)$$

$$R_2/Z_0 = \left[-1.78 + 0.749\frac{w}{h}\right]\frac{S}{h} + \left[1.196 - 0.971\ln\left(\frac{w}{h}\right)\right]\sinh\left(2.3345\frac{S}{h}\right),$$

$$(3.110)$$

where Z_0 is the characteristic impedance of the microstrip line calculated using the equations presented in Section 3.15.

The S-parameters of the gap discontinuity can then be obtained using the equations for a 'π' admittance network in Table 1.8, Section 1.5, i.e.

$$[S] = \frac{\begin{bmatrix} Y_{0,\text{ref}}^2 - (Y_1 - Y_2)Y_{0,\text{ref}} - (Y_1 Y_2 + Y_2 Y_3 + Y_3 Y_1) & 2Y_{0,\text{ref}} Y_3 \\ 2Y_{0,\text{ref}} Y_3 & Y_{0,\text{ref}}^2 - (Y_2 - Y_1)Y_{0,\text{ref}} - (Y_1 Y_2 + Y_2 Y_3 + Y_3 Y_1) \end{bmatrix}}{Y_{0,\text{ref}}^2 + (Y_1 + Y_2 + 2Y_3)Y_{0,\text{ref}} + Y_1 Y_2 + Y_2 Y_3 + Y_3 Y_1}$$

$$(3.111)$$

with

$$Y_1 = Y_2 = \left(j\omega L_{11} + \frac{1}{j\omega C_{11}} \right)^{-1} + \left(R_1 + j\omega L_{12} + \frac{1}{j\omega C_{12}} \right)^{-1} \quad \text{and}$$

$$Y_3 = \left(R_2 + j\omega L_2 + \frac{1}{j\omega C_2} \right)^{-1}. \tag{3.112}$$

3.20.3 Bend Discontinuity

The 90° bend discontinuity can also be represented by series inductance L_1, shunt capacitance C_s and series inductance L_2 (Wadell, 1991), where

$$L_1 = L_2 = \frac{L_s}{2} = L_s = 50h \left(4\sqrt{\frac{w}{h}} - 4.21 \right) \text{ (nH)},$$

$$C_s = \begin{cases} \left[\dfrac{(14\varepsilon_r + 12.5)\,(w/h) - (1.83\varepsilon_r - 2.25)}{\sqrt{w/h}} + \dfrac{0.02\varepsilon_r}{w/h} \right] w \text{ (pF)} & \text{if } w < h \\[3mm] \left[(9.5\varepsilon_r + 1.25) \left(\dfrac{w}{h} \right) + 5.2\varepsilon_r + 7 \right] w \text{ (pF)} & \text{if } w \geq h \end{cases} \tag{3.113}$$

These formulas are accurate to within 5% for $2.5 \leq \varepsilon_r \leq 15.0$ and $0.1 \leq w/h \leq 5.0$.

The S-parameters of the bend discontinuity can then be obtained using the equations for a 'T' impedance network in Table 1.8, Section 1.5, i.e.

$$[S] = \frac{\begin{bmatrix} -Z_{0,\text{ref}}^2 + (Z_1 - Z_2)Z_{0,\text{ref}} + Z_1 Z_2 + Z_2 Z_3 + Z_3 Z_1 & 2Z_{0,\text{ref}} Z_3 \\ 2Z_{0,\text{ref}} Z_3 & -Z_{0,\text{ref}}^2 + (Z_2 - Z_1)Z_{0,\text{ref}} + Z_1 Z_2 + Z_2 Z_3 + Z_3 Z_1 \end{bmatrix}}{Z_{0,\text{ref}}^2 + (Z_1 + Z_2 + 2Z_3)Z_{0,\text{ref}} + Z_1 Z_2 + Z_2 Z_3 + Z_3 Z_1} \tag{3.114}$$

with

$$Z_1 = j\omega L_1, \quad Z_2 = j\omega L_2 \quad \text{and} \quad Z_3 = 1/(j\omega C_s). \tag{3.115}$$

3.20.4   Slit Discontinuity

The slit discontinuity can be modelled by series inductance L_N given by (Gupta, Garg, Bahl and Bhartia, 1996)

$$L_N = 2000h \left(1 - \frac{Z_{0m}}{Z'_{0m}} \sqrt{\frac{\varepsilon_{re}}{\varepsilon'_{re}}} \right)^2 \text{(nH)}, \qquad (3.116)$$

where Z_{0m} and ε_{re} are the characteristic impedance and the effective relative dielectric constant of the microstrip line with width w, respectively, and Z'_{0m} and ε'_{re} are the values for the microstrip line with width $(w-b)$.

The S-parameters of the slit discontinuity can then be obtained using the equations for a two-port single series impedance described in Section 3.4, i.e.

$$[S] = \begin{bmatrix} \dfrac{Z}{Z + 2Z_{0,ref}} & \dfrac{2Z_{0,ref}}{Z + 2Z_{0,ref}} \\ \dfrac{2Z_{0,ref}}{Z + 2Z_{0,ref}} & \dfrac{Z}{Z + 2Z_{0,ref}} \end{bmatrix} \qquad (3.117)$$

with

$$Z = j\omega L_N. \qquad (3.118)$$

3.20.5   Open-End Discontinuity

The open-end discontinuity can be modelled as an equivalent admittance given by (Alexopolous and Wu, 1994)

$$Y = j\omega C_1 + \left(R + j\omega L + \frac{1}{j\omega C_2} \right)^{-1} \qquad (3.119)$$

with

$$C_1 \times \frac{25 Z_0}{h} = 1.125 \tanh \left(1.358 \frac{w}{h} \right) - 0.315 \,(\text{pF}\,\Omega),$$

$$C_2 \times \frac{25 Z_0}{h} = 6.832 \tanh \left(0.0109 \frac{w}{h} \right) + 0.910 \,(\text{pF}\,\Omega),$$

$$L \times \frac{25}{h Z_0} = 0.008285 \tanh \left(0.5665 \frac{w}{h} \right) + 0.0103 \,(\text{pF}\,\Omega),$$

$$R/Z_0 = 1.024 \tanh\left(2.025\frac{w}{b}\right), \tag{3.120}$$

where Z_0 is the characteristic impedance of the microstrip line calculated using the equations in Section 3.15.

The S-parameter of the open-end discontinuity can then be obtained using the equations for a one-port terminated load described in Section 3.3, i.e.

$$S_{11} = \frac{Z_L - Z_{0,\mathrm{ref}}}{Z_L + Z_{0,\mathrm{ref}}} \tag{3.121}$$

with

$$Z_L = 1/Y. \tag{3.122}$$

3.21 PHYSICAL LINE DISCONTINUITIES: STRIPLINE DISCONTINUITIES

Figure 3.21(a)–(d) shows, respectively, the stripline step, gap, bend and open-end discontinuities together with their equivalent circuits. The models of these discontinuities are described below.

3.21.1 Step Discontinuity

The step discontinuity can be modelled simply by series inductance L_s given by (Gupta, Garg and Chadha, 1981)

$$L_s = \frac{\sqrt{\varepsilon_r}\,w_1' Z_{01}}{\pi c} \ln\left(\frac{1}{\sin\left(\pi w_2'/2w_1'\right)}\right), \tag{3.123}$$

where

$$Z_{01} = Z_0(w = w_1), \quad w_1' = w'(w = w_1), \quad w_2' = w'(w = w_2) \tag{3.124}$$

with (Pozar, 1990; Howe, 1974)

$$Z_0 = \frac{30\pi b}{\sqrt{\varepsilon_r}\,(w_e + 0.441b)},$$

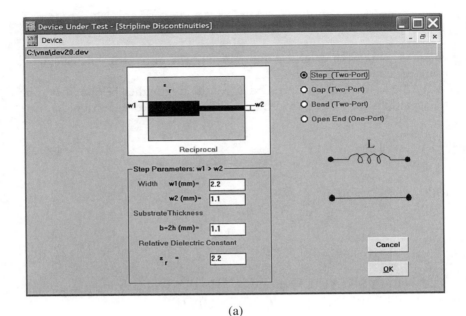

(a)

(b)

Figure 3.21 Physical Line Discontinuities: Stripline Discontinuities: (a) step, (b) gap, (c) bend and (d) open end

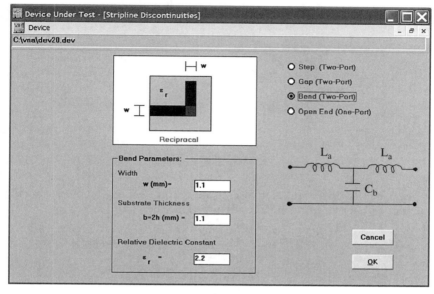

(c)

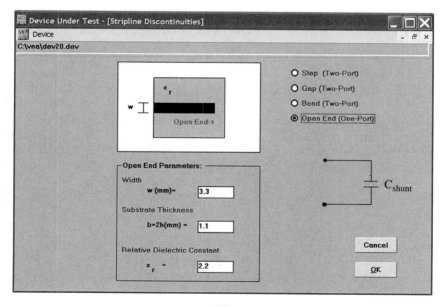

(d)

Figure 3.21 (Continued)

$$w_e = \begin{cases} w & \text{if } w > 0.35h \\ w - h\left(0.35 - \frac{w}{h}\right)^2 & \text{if } w \le 0.35h \end{cases},$$

$$w = \begin{cases} h\dfrac{K(k)}{K'(k)} & \text{if } w < 0.5h \\ w - h\left(0.35 - \dfrac{w}{h}\right)^2 & \text{if } w > 0.5h \end{cases},$$

$$k = \tanh\left(\frac{\pi w}{2h}\right). \tag{3.125}$$

The S-parameters of the step discontinuity can then be obtained using the equations for a two-port single series impedance described in Section 3.4, i.e.

$$[S] = \begin{bmatrix} \dfrac{Z}{Z+2Z_{0,\text{ref}}} & \dfrac{2Z_{0,\text{ref}}}{Z+2Z_{0,\text{ref}}} \\ \dfrac{2Z_{0,\text{ref}}}{Z+2Z_{0,\text{ref}}} & \dfrac{Z}{Z+2Z_{0,\text{ref}}} \end{bmatrix} \tag{3.126}$$

with

$$Z = j\omega L_s. \tag{3.127}$$

3.21.2   Gap Discontinuity

The gap discontinuity can be modelled by a 'π' network of three capacitors (Gupta, Garg and Chadha, 1981), as shown in Figure 3.21(b). The susceptance values of these three capacitors normalised to the characteristic admittance of the stripline are, respectively,

$$\frac{\omega C_1}{Y_o} = \overline{B}_1 = \frac{1 + \overline{B}_a \cot(\beta S/2)}{\cot(\beta S/2) - \overline{B}_a},$$

$$\frac{\omega C_{12}}{Y_o} = \overline{B}_{12} = \frac{1}{2}\left(\frac{1 + \left(2\overline{B}_b + \overline{B}_a\right)\cot(\beta S/2)}{\cot(\beta S/2) - \left(2\overline{B}_b + \overline{B}_a\right)} - \overline{B}_1 \right), \tag{3.128}$$

where

$$\overline{B}_a = -\frac{2b}{\lambda}\ln\cosh(\pi S/2b), \quad \overline{B}_b = \frac{b}{\lambda}\ln\coth(\pi S/2b). \tag{3.129}$$

The S-parameters of the gap discontinuity can then be obtained using the equations for a 'π' admittance network in Table 1.8, Section 1.5, i.e.

$$[S] = \frac{\begin{bmatrix} Y_{0,\text{ref}}^2 - (Y_1 - Y_2)Y_{0,\text{ref}} - (Y_1 Y_2 + Y_2 Y_3 + Y_3 Y_1) & 2Y_{0,\text{ref}} Y_3 \\ 2Y_{0,\text{ref}} Y_3 & Y_{0,\text{ref}}^2 - (Y_2 - Y_1)Y_{0,\text{ref}} - (Y_1 Y_2 + Y_2 Y_3 + Y_3 Y_1) \end{bmatrix}}{Y_{0,\text{ref}}^2 + (Y_1 + Y_2 + 2Y_3)Y_{0,\text{ref}} + Y_1 Y_2 + Y_2 Y_3 + Y_3 Y_1}$$

(3.130)

with

$$Y_1 = Y_2 = j\omega C_1 \quad \text{and} \quad Y_3 = j\omega C_{12}.$$

(3.131)

3.21.3   Bend Discontinuity

The 90° bend discontinuity can be modelled by series inductance L_1, shunt capacitance C_s and series inductance $L_2 = L_1$ (Gupta, Garg and Chadha, 1981). Their normalised reactance values to the characteristic impedance are given by

$$\frac{\omega L_1}{Z_0} = \frac{\omega L_2}{Z_0} = \overline{X}_a = \frac{w'}{\lambda}[1.756 + 4(w'/\lambda)^2],$$

$$\frac{-1}{\omega C_s Z_0} = \overline{X}_b = \frac{w'}{\lambda}[0.0725 - 0.159(\lambda/w')^2],$$

(3.132)

with

$$w' = \begin{cases} h\dfrac{K(k)}{K'(k)} & \text{if } w < 0.5h \\ w + \dfrac{2h}{\pi}\ln 2 & \text{if } w > 0.5h \end{cases} \quad \text{and} \quad k = \tanh\left(\frac{\pi w}{2h}\right).$$

(3.133)

The S-parameters of the bend discontinuity can then be obtained using the equations for a 'T' impedance network in Table 1.8, Section 1.5, i.e.

$$[S] = \frac{\begin{bmatrix} -Z_{0,\text{ref}}^2 + (Z_1 - Z_2)Z_{0,\text{ref}} + Z_1 Z_2 + Z_2 Z_3 + Z_3 Z_1 & 2Z_{0,\text{ref}} Z_3 \\ 2Z_{0,\text{ref}} Z_3 & -Z_{0,\text{ref}}^2 + (Z_2 - Z_1)Z_{0,\text{ref}} + Z_1 Z_2 + Z_2 Z_3 + Z_3 Z_1 \end{bmatrix}}{Z_{0,\text{ref}}^2 + (Z_1 + Z_2 + 2Z_3)Z_{0,\text{ref}} + Z_1 Z_2 + Z_2 Z_3 + Z_3 Z_1}$$

(3.134)

with

$$Z_1 = j\omega L_1, \quad Z_2 = j\omega L_2 \quad \text{and} \quad Z_3 = 1/(j\omega C_s).$$

(3.135)

3.21.4   Open-End Discontinuity

The open end discontinuity can be modelled as an equivalent shunt capacitor of capacitance (Gupta, Garg and Chadha, 1981)

$$C_{\text{shunt}} = \frac{\sqrt{\varepsilon_r}\Delta l}{cZ_0}, \tag{3.136}$$

where Z_0 is the characteristic impedance of the strip line and

$$\beta\Delta l = \tan^{-1}\left[\frac{\delta+2W}{4\delta+2W}\tan(\beta\delta)\right], \quad \delta = h\frac{\ln 2}{\pi}, \quad \beta = \frac{2\pi}{\lambda} \quad \text{and} \quad \lambda = \frac{\lambda_0}{\sqrt{\varepsilon_r}}. \tag{3.137}$$

The S-parameter of the open-end discontinuity can then be obtained using the equations for a one-port terminated load described in Section 3.3, i.e.

$$S_{11} = \frac{Z_L - Z_{0,\text{ref}}}{Z_L + Z_{0,\text{ref}}} \tag{3.138}$$

with

$$Z_L = 1/(j\omega C_{\text{shunt}}). \tag{3.139}$$

3.22   GENERAL COUPLED LINES: FOUR-PORT COUPLED LINES

Figure 3.22 shows the display of the General Coupled Lines: Four-Port Coupled Lines device builder. The coupled line can be modelled using the even–odd mode analysis (Pozar, 1990; Napoli and Hughes, 1970), which leads to the S-parameters for the four-port device:

$$[S] = \begin{bmatrix} S_{11} & S_{21} & S_{31} & S_{41} \\ S_{21} & S_{11} & S_{41} & S_{31} \\ S_{31} & S_{41} & S_{11} & S_{21} \\ S_{41} & S_{31} & S_{21} & S_{11} \end{bmatrix}. \tag{3.140}$$

For a lossless coupled line,

$$S_{\substack{11 \\ 31}} = \frac{\rho_e}{2}\left(1 - \frac{(1-\rho_e^2)}{(e^{j2\beta_e L} - \rho_e^2)}\right) \pm \frac{\rho_o}{2}\left(1 - \frac{(1-\rho_o^2)\,e^{j\beta_o L}}{(e^{j2\beta_o L} - \rho_o^2)}\right) \quad \text{and}$$

$$S_{\substack{21 \\ 41}} = \left(\frac{(1-\rho_e^2)\,e^{j\beta_e L}}{2\,(e^{j2\beta_e L} - \rho_e^2)}\right) \pm \left(\frac{(1-\rho_o^2)\,e^{j\beta_o L}}{2\,(e^{j2\beta_o L} - \rho_o^2)}\right), \tag{3.141}$$

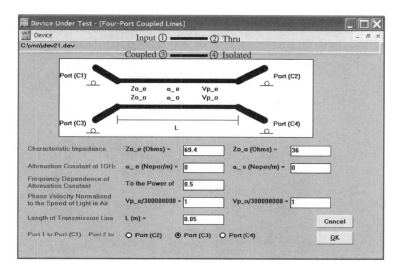

Figure 3.22 General Coupled Lines: Four-Port Coupled Lines

where

$$\rho_e = -\frac{(1 - Z_{0e}/Z_0)}{(1 + Z_{0e}/Z_0)}, \quad \rho_o = -\frac{(1 - Z_{0o}/Z_0)}{(1 + Z_{0o}/Z_0)}, \qquad (3.142)$$

Z_{0e} is the characteristic impedance of the even mode, Z_{0o} the characteristic impedance of the odd mode $(Z_{0o} < Z_{0e})$, β_e the phase constant of the even mode and β_o the phase constant of the odd mode $(\beta_o > \beta_e)$, and $Z_0 = Z_{0,\text{ref}}$.

If the coupled line is not lossless, the parameters $j\beta_e$ and $j\beta_o$ in the above equations are replaced, respectively, by

$$\gamma_e = \alpha_e + j\beta_e \quad \text{and} \quad \gamma_o = \alpha_o + j\beta_o. \qquad (3.143)$$

3.23 GENERAL COUPLED LINES: TWO-PORT COUPLED LINES

Figure 3.23 shows the display of the General Coupled Lines: Two-Port Coupled Lines device builder. Port 1 of the device is connected to Port 1 of the VNA, and Port 4 to Port 2 of the VNA. Ports 2 and 3 of the device can be terminated by a short-circuit or open-circuit load with a reflection coefficient of $\Gamma_2 = \Gamma_3 = -1$ or $\Gamma_2 = \Gamma_3 = 1$. Since

$$
\begin{bmatrix} b_1 \\ b_2 \\ b_3 \\ b_4 \end{bmatrix} =
\begin{bmatrix}
S_{11} & S_{21} & S_{31} & S_{41} \\
S_{21} & S_{11} & S_{41} & S_{31} \\
S_{31} & S_{41} & S_{11} & S_{21} \\
S_{41} & S_{31} & S_{21} & S_{11}
\end{bmatrix}
\begin{bmatrix} a_1 \\ a_2 \\ a_3 \\ a_4 \end{bmatrix}
\qquad (3.144)
$$

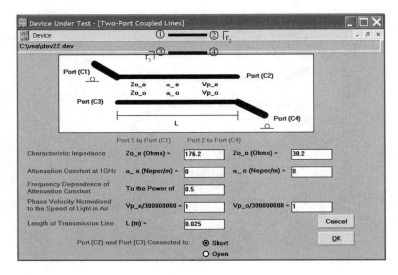

Figure 3.23   General Coupled Lines: Two-Port Coupled Lines

and

$$
\begin{bmatrix} b_1 \\ b_4 \end{bmatrix} = \begin{bmatrix} S_{11} & S_{41} \\ S_{41} & S_{11} \end{bmatrix}\begin{bmatrix} a_1 \\ a_4 \end{bmatrix} + \begin{bmatrix} S_{21} & S_{31} \\ S_{31} & S_{21} \end{bmatrix}\begin{bmatrix} \Gamma_2 & 0 \\ 0 & \Gamma_3 \end{bmatrix}\begin{bmatrix} 1 - \Gamma_2 S_{11} & -\Gamma_3 S_{41} \\ -\Gamma_2 S_{41} & 1 - \Gamma_3 S_{11} \end{bmatrix}^{-1}
$$
$$
\begin{bmatrix} S_{21} & S_{31} \\ S_{31} & S_{21} \end{bmatrix}\begin{bmatrix} a_1 \\ a_4 \end{bmatrix},
$$

$$
\begin{bmatrix} b_1 \\ b_4 \end{bmatrix} = \begin{bmatrix} S'_{11} & S'_{41} \\ S'_{41} & S'_{11} \end{bmatrix}\begin{bmatrix} a_1 \\ a_4 \end{bmatrix} = \begin{bmatrix} S''_{11} & S''_{12} \\ S''_{21} & S''_{22} \end{bmatrix}\begin{bmatrix} a_1 \\ a_4 \end{bmatrix}, \tag{3.145}
$$

the new two-port S-parameters are

$$
\begin{bmatrix} S''_{11} & S''_{12} \\ S''_{21} & S''_{22} \end{bmatrix} = \begin{bmatrix} S_{11} & S_{41} \\ S_{41} & S_{11} \end{bmatrix} + \begin{bmatrix} S_{21} & S_{31} \\ S_{31} & S_{21} \end{bmatrix}\begin{bmatrix} \Gamma_2 & 0 \\ 0 & \Gamma_3 \end{bmatrix}\begin{bmatrix} 1 - \Gamma_2 S_{11} & -\Gamma_3 S_{41} \\ -\Gamma_2 S_{41} & 1 - \Gamma_3 S_{11} \end{bmatrix}^{-1}
$$
$$
\begin{bmatrix} S_{21} & S_{31} \\ S_{31} & S_{21} \end{bmatrix},
$$
$$
\tag{3.146}
$$

where S_{11}, S_{21}, S_{31} and S_{41} are given in Section 3.22.

3.24 PHYSICAL COUPLED LINES: FOUR-PORT COUPLED MICROSTRIP LINES

Figure 3.24 shows the display of the Physical Coupled Lines: Four-Port Coupled Microstrip Lines device builder. The coupled line can also be modelled using the even–odd mode analysis.

The effective relative dielectric constant of the even mode is given by (Gupta, Garg, Bahl and Bhartia, 1996)

$$\varepsilon_{re}^{e}(0) = \frac{\varepsilon_r + 1}{2} + \frac{\varepsilon_r - 1}{2} \left(1 + \frac{10}{v}\right)^{-a_e b_e},$$

$$v = \frac{u\left(20 + g^2\right)}{10 + g^2} + ge^{-g},$$

$$a_e = 1 + \frac{1}{49} \ln\left[\frac{v^4 + \left(v/52\right)^2}{v^4 + 0.432}\right] + \frac{1}{18.7} \ln\left[1 + \left(\frac{v}{18.1}\right)^3\right],$$

$$b_e = 0.564 \left(\frac{\varepsilon_r - 0.9}{\varepsilon_r + 3}\right)^{0.053}. \tag{3.147}$$

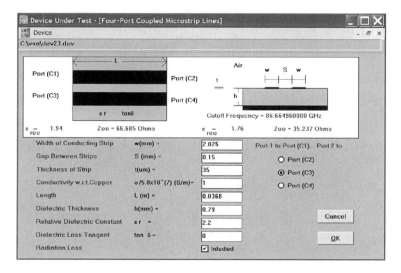

Figure 3.24 Physical Coupled Lines: Four-Port Coupled Microstrip Lines

This equation has an error of 0.7% for $0.1 \le u \le 10$, $0.1 \le g \le 10$ and $1 \le \varepsilon_r \le 18$, where $u = w/h$ and $g = S/h$.

The effective relative dielectric constant of the odd mode is given by

$$\varepsilon_{re}^o(0) = \varepsilon_{re}(0) + \{0.5(\varepsilon_r + 1) - \varepsilon_{re}(0) + a_o\}e^{-c_o(g^{d_o})}, \qquad (3.148)$$

where $\varepsilon_{re}(0)$ is the effective relative dielectric constant of a single microstrip line,

$$a_o = 0.7287(\varepsilon_{re}(0) - 0.5(\varepsilon_r + 1))(1 - e^{-0.179u}),$$

$$b_o = \frac{0.747\varepsilon_r}{0.15 + \varepsilon_r},$$

$$c_o = b_o - (b_o - 0.207)e^{-0.414u},$$

$$d_o = 0.593 + 0.694e^{-0.562u}. \qquad (3.149)$$

The characteristic impedances of the even and odd modes are, respectively,

$$Z_{0e} = Z_0 \frac{\sqrt{\varepsilon_{re}(0)/\varepsilon_{re}^e(0)}}{1 - Q_4\sqrt{\varepsilon_{re}(0)}(Z_0/377)} \quad \text{and} \quad Z_{0o} = Z_0 \frac{\sqrt{\varepsilon_{re}(0)/\varepsilon_{re}^o(0)}}{1 - Q_{10}\sqrt{\varepsilon_{re}(0)}(Z_0/377)}, \qquad (3.150)$$

where Z_0 is the characteristic impedance of a single microstrip line calculated using the equations in Section 3.15, and

$$Q_1 = 0.88695u^{0.194},$$

$$Q_2 = 1 + 0.7519g + 0.189g^{2.31},$$

$$Q_3 = 0.1975 + \left[16.6 + \left(\frac{8.4}{g}\right)^6\right]^{-0.387} + \frac{1}{241}\ln\left[\frac{g^{10}}{1 + (g/3.4)^{10}}\right],$$

$$Q_4 = \frac{2Q_1}{Q_2}\frac{1}{u^{Q_3}e^{-g} + (2 - e^{-g})u^{-Q_3}},$$

$$Q_5 = 1.974 + 1.14\ln\left[1 + \frac{0.638}{g + 0.517g^{2.43}}\right],$$

$$Q_6 = 0.2305 + \frac{1}{281.3}\ln\left[\frac{g^{10}}{1 + (g/5.8)^{10}}\right] + \frac{1}{5.1}\ln(1 + 0.598g^{1.154}),$$

$$Q_7 = \frac{10 + 190g^2}{1 + 82.3g^3},$$

$$Q_8 = e^{-[6.5 + 0.95 \ln(g) + (g/0.15)^5]},$$

$$Q_9 = \ln(Q_7)(Q_8 + 1/16.5),$$

$$Q_{10} = Q_4 - \frac{Q_5}{Q_2} e^{[Q_6 \ln(u)/u^{Q_9}]}. \tag{3.151}$$

The attenuation constant due to the conductor loss can be calculated using the Incremental Inductance Rule (Pozar, 1990). The attenuation constants due to the conductor loss for the even and odd modes are, respectively,

$$\alpha_c^e = \sum_{\Omega} \frac{R_s}{240\pi Z_{0e}} \frac{\partial \left(\sqrt{\varepsilon_{re}^o} Z_{0e}\right)}{\partial \Omega}, \quad \alpha_c^o = \sum_{\Omega} \frac{R_s}{240\pi Z_{0o}} \frac{\partial \left(\sqrt{\varepsilon_{re}^o} Z_{0o}\right)}{\partial \Omega}, \tag{3.152}$$

where $R_s = \sqrt{\omega\mu_0/2\sigma}$ is the surface resistance of the microstrip line, and Ω represents a dimensional parameter of the microstrip coupled line.

The attenuation constants due to dielectric losses for the even and odd modes, respectively, are given by

$$\alpha_d^{e,o} = \frac{\varepsilon_r}{\sqrt{\varepsilon_{re}^{e,o}}} \frac{\varepsilon_{re}^{e,o} - 1}{\varepsilon_r - 1} \frac{\tan\delta}{\lambda_0} \ (\text{Neper/m}). \tag{3.153}$$

The radiation loss of the odd mode is negligible. The attenuation constant of the even mode due to radiation loss is given by

$$\alpha_r^e = \frac{\pi f}{c Q_r} = \frac{\pi f}{c} \left(\frac{480\pi (hf/c)^2 R \sqrt{\varepsilon_{re}^e}}{Z_{0e}} \right). \tag{3.154}$$

with

$$R = \frac{\varepsilon_{re}^e + 1}{\varepsilon_{re}^e} - \frac{(\varepsilon_{re}^e - 1)^2}{2\varepsilon_{re}^{e3/2}} \ln \frac{\sqrt{\varepsilon_{re}^e} + 1}{\sqrt{\varepsilon_{re}^e} - 1}. \tag{3.155}$$

Hence the coupled microstrip line has the following parameters:

$$\gamma_e = \alpha_e + j\beta_e \quad \text{and} \quad \gamma_o = \alpha_o + j\beta_o, \tag{3.156}$$

where

$$\alpha_e = \alpha_c^e + \alpha_d^e + \alpha_r^e, \quad \alpha_o = \alpha_c^o + \alpha_d^o, \quad \beta_e = \frac{2\pi f}{c} \sqrt{\varepsilon_{re}^e} \quad \text{and} \quad \beta_o = \frac{2\pi f}{c} \sqrt{\varepsilon_{re}^o}. \tag{3.157}$$

The S-parameters of the four-port coupled microstrip line are therefore

$$[S] = \begin{bmatrix} S_{11} & S_{21} & S_{31} & S_{41} \\ S_{21} & S_{11} & S_{41} & S_{31} \\ S_{31} & S_{41} & S_{11} & S_{21} \\ S_{41} & S_{31} & S_{21} & S_{11} \end{bmatrix} \qquad (3.158)$$

with

$$S_{\underset{31}{11}} = \frac{\rho_e}{2}\left(1 - \frac{(1-\rho_e^2)}{(e^{2\gamma_e L}-\rho_e^2)}\right) \pm \frac{\rho_o}{2}\left(1 - \frac{(1-\rho_o^2)\,e^{\gamma_e L}}{(e^{2\gamma_e L}-\rho_o^2)}\right) \quad \text{and}$$

$$S_{\underset{41}{21}} = \left(\frac{(1-\rho_e^2)\,e^{\gamma_e L}}{2\,(e^{2\gamma_e L}-\rho_e^2)}\right) \pm \left(\frac{(1-\rho_o^2)\,e^{\gamma_e L}}{2\,(e^{2\gamma_e L}-\rho_o^2)}\right), \qquad (3.159)$$

where

$$\rho_e = -\frac{(1-Z_{0e}/Z_0)}{(1+Z_{0e}/Z_0)}, \quad \rho_o = -\frac{(1-Z_{0o}/Z_0)}{(1+Z_{0o}/Z_0)} \quad \text{and} \quad Z_0 = Z_{0,\text{ref}}. \quad (3.160)$$

3.25   PHYSICAL COUPLED LINES: TWO-PORT COUPLED MICROSTRIP LINES

Figure 3.25 shows the display of the Physical Coupled Lines: Two-Port Coupled Microstrip Lines device builder. Port 1 of the device is connected

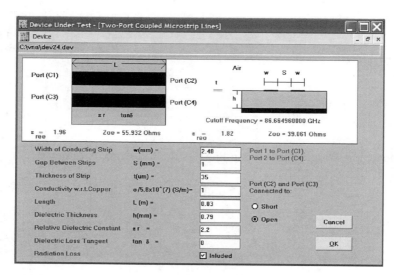

Figure 3.25   Physical Coupled Lines: Two-Port Coupled Microstrip Lines

to Port 1 of the VNA, and Port 4 to Port 2 of the VNA. Ports 2 and 3 of the device can be terminated by a short-circuit or open-circuit load with a reflection coefficient of $\Gamma_2 = \Gamma_3 = -1$ or $\Gamma_2 = \Gamma_3 = 1$. As shown in Section 3.23, the S-parameters of the terminated coupled microstrip line are

$$\begin{bmatrix} S_{11}'' & S_{12}'' \\ S_{21}'' & S_{22}'' \end{bmatrix} = \begin{bmatrix} S_{11} & S_{41} \\ S_{41} & S_{11} \end{bmatrix} + \begin{bmatrix} S_{21} & S_{31} \\ S_{31} & S_{21} \end{bmatrix} \begin{bmatrix} \Gamma_2 & 0 \\ 0 & \Gamma_3 \end{bmatrix} \begin{bmatrix} 1 - \Gamma_2 S_{11} & -\Gamma_3 S_{41} \\ -\Gamma_2 S_{41} & 1 - \Gamma_3 S_{11} \end{bmatrix}^{-1} \begin{bmatrix} S_{21} & S_{31} \\ S_{31} & S_{21} \end{bmatrix},$$

(3.161)

where S_{11}, S_{21}, S_{31} and S_{41} are given in Section 3.24.

3.26 LUMPED ELEMENTS: INDUCTORS

Figure 3.26(a)–(c) shows circular coil, circular spiral and single turn lumped inductor device builders and their equivalent circuits, respectively. These lumped elements models are described below.

3.26.1 Circular Coil

The circular coil can be modelled as an inductor L in series with a resistor R and in parallel with a capacitor C (Wadell, 1991). The overall impedance of the coil is

$$Z = \frac{(R + j\omega L)(1/j\omega C)}{(R + j\omega L) + (1/j\omega C)}.$$

(3.162)

The inductance L is

$$L = \begin{pmatrix} \dfrac{\mu_0 n^2 \pi a^2}{b} \left[f_1\left(\dfrac{4a^2}{b^2}\right) - \dfrac{4}{3\pi} \dfrac{2a}{b} \right] & b \geq 2a \\[4mm] \mu_0 n^2 a^2 \left\{ \left[\ln\left(\dfrac{8a}{b}\right) - \dfrac{1}{2} \right] f_1\left(\dfrac{b^2}{4a^2}\right) + f_2\left(\dfrac{b^2}{4a^2}\right) \right\} & b < 2a \end{pmatrix},$$

(3.163)

where

$$f_1(x) = \frac{1 + 0.383901x + 0.017108x^2}{1 + 0.258952x} \quad \text{and}$$

$$f_2(x) = 0.093842x + 0.002029x^2 - 0.000801x^3$$

(3.164)

with $0 \leq x \leq 1.0$. The resistance R is

$$
R = \begin{cases} \dfrac{R_s}{2\pi a_w}(2\pi an) = \dfrac{R_s an}{a_w}\ (\Omega) & \text{if } \delta_s < a_w \\[2ex] \dfrac{1}{6}\dfrac{(2\pi an)}{\pi a_w^2}\ (\Omega) & \text{if } \delta_s > a_w \end{cases}, \tag{3.165}
$$

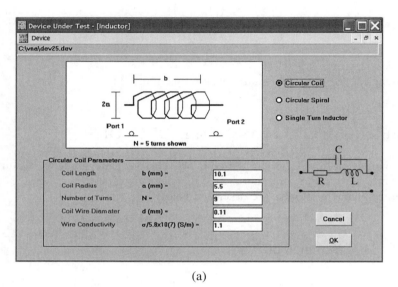

(a)

(b)

Figure 3.26 Lumped Elements: Inductors: (a) circular coil, (b) circular spiral and (c) single turn inductor

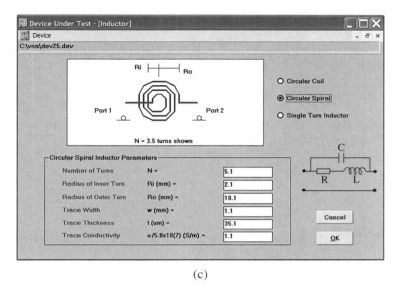

(c)

Figure 3.26 (Continued)

where R_s is the surface resistance of the coil, δ_s the skin depth and $a_w = d/2$. The stray capacitance C is given by (Medhurst, 1947)

$$C = 200aH \, (\text{pF}), \tag{3.166a}$$

where

$$H = 0.1126\frac{b}{2a} + 0.08 + \frac{0.27}{\sqrt{b/2a}}. \tag{3.166b}$$

The S-parameters of the circular coil in two-port connection can then be obtained using the equations for a two-port single series impedance described in Section 3.4, i.e.

$$[S] = \begin{bmatrix} \dfrac{Z}{Z + 2Z_{0,\text{ref}}} & \dfrac{2Z_{0,\text{ref}}}{Z + 2Z_{0,\text{ref}}} \\[2mm] \dfrac{2Z_{0,\text{ref}}}{Z + 2Z_{0,\text{ref}}} & \dfrac{Z}{Z + 2Z_{0,\text{ref}}} \end{bmatrix}. \tag{3.167}$$

3.26.2 Circular Spiral

The circular spiral lumped inductor can be modelled in the same equivalent circuit as the circular coil (Wadell, 1991; Gupta, Garg and Chadha, 1981).

However, the inductance is given by

$$L = \frac{n^2 a^2}{8a + 11c_{oi}} \text{ (nH)}, \tag{3.168}$$

where $a = (r_o + r_i)/2$, $c_{oi} = r_o - r_i$ which are in mils or

$$a_{mil} = \left(\frac{1000}{25.4}\right) a_{mm} = \frac{10^6 a_{meter}}{25.4}, \quad c_{oi,mil} = \left(\frac{1000}{25.4}\right) c_{oi,mm} = \frac{10^6 c_{oi,meter}}{25.4}. \tag{3.169}$$

The stray capacitance is given by

$$C = 0.035 \, (2r_{0mm}) + 0.06 \, \text{(pF)} \tag{3.170}$$

and the resistance R by

$$R = \begin{cases} R_s \dfrac{2\pi \, (0.5 \, (r_o + r_i)) \, n}{2 \, (w + t)} = \dfrac{R_s an}{a_w} & \text{if } \delta_s < \dfrac{1}{2} \dfrac{wt}{w + t} \\[2mm] \dfrac{1}{\sigma} \dfrac{2\pi \, (0.5 \, (r_o + r_i) \, n)}{wt} & \text{if } \delta_s > \dfrac{1}{2} \dfrac{wt}{w + t} \end{cases}, \tag{3.171}$$

where R_s is the surface resistance of the coil and δ_s the skin depth.

The S-parameters of the circular spiral in two-port connection are then given by

$$[S] = \begin{bmatrix} \dfrac{Z}{Z + 2Z_{0,ref}} & \dfrac{2Z_{0,ref}}{Z + 2Z_{0,ref}} \\[3mm] \dfrac{2Z_{0,ref}}{Z + 2Z_{0,ref}} & \dfrac{Z}{Z + 2Z_{0,ref}} \end{bmatrix}, \tag{3.172}$$

where

$$Z = \frac{(R + j\omega L) \, (1/j\omega C)}{(R + j\omega L) + (1/j\omega C)}. \tag{3.173}$$

3.26.3   Single Turn Inductor

The single turn inductor can be modelled as an inductor L in series with a resistor R (Wadell, 1991). The inductance value is given by

$$L = 0.02339 \left[(S_1 + S_2) \ln\left(\frac{2S_1 S_2}{w + t}\right) - S_1 \ln\,(S_1 + g) - S_2 \ln\,(S_2 + g) \right]$$

$$+ 0.01016 \left[2g - \frac{S_1 + S_2}{2} + 0.447 \, (w + t) \right] (\mu H), \tag{3.174}$$

where

$$g = \sqrt{S_1^2 + S_2^2} \qquad (3.175)$$

and S_1 and S_2 are in inches which can be converted from millimetre or metre using

$$S_{1,2} = \frac{S_{1,2mm}}{25.4} = \frac{S_{1,2m} \times 10^3}{25.4} \text{ inch.} \qquad (3.176)$$

The resistance R is given by

$$R = \begin{cases} R_s \dfrac{2\,(S_1 + S_2)}{2\,(w + t)} & \text{if } \delta_s < \dfrac{1}{2}\dfrac{wt}{w+t} \\[2ex] \dfrac{1}{\sigma}\dfrac{2(S_1+S_2)}{wt} & \text{otherwise} \end{cases} . \qquad (3.177)$$

The S-parameters of the single turn inductor in two-port connection are then given by

$$[S] = \begin{bmatrix} \dfrac{Z}{Z + 2Z_{0,\text{ref}}} & \dfrac{2Z_{0,\text{ref}}}{Z + 2Z_{0,\text{ref}}} \\[3ex] \dfrac{2Z_{0,\text{ref}}}{Z + 2Z_{0,\text{ref}}} & \dfrac{Z}{Z + 2Z_{0,\text{ref}}} \end{bmatrix}, \qquad (3.178)$$

where

$$Z = R + j\omega L. \qquad (3.179)$$

3.27 LUMPED ELEMENTS: CAPACITORS

Figure 3.27(a) and (b) shows, respectively, the thin film and interdigital lumped capacitor device builders and their equivalent circuits. The parameters of the equivalent circuits are described below.

3.27.1 Thin Film Capacitor

The thin film capacitor is modelled as a capacitor with capacitance (Wadell, 1991)

$$C = \frac{\varepsilon_r \varepsilon_0}{b}\left[\left(l + \frac{4b\ln 2}{\pi}\right)\left(W + \frac{4b\ln 2}{\pi}\right)\right] \text{ (F).} \qquad (3.180)$$

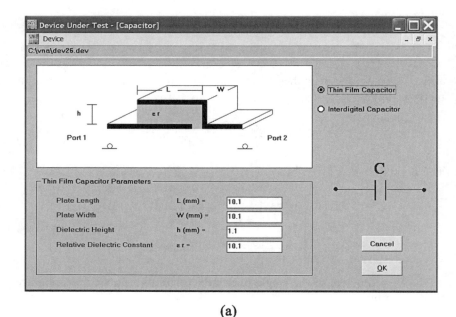

(a)

(b)

Figure 3.27 Lumped Elements: Capacitors: (a) thin film capacitor and (b) interdigital capacitor

Hence the S-parameters of the thin film capacitor in two-port connection are given by

$$[S] = \begin{bmatrix} \dfrac{Z}{Z+2Z_{0,\text{ref}}} & \dfrac{2Z_{0,\text{ref}}}{Z+2Z_{0,\text{ref}}} \\ \dfrac{2Z_{0,\text{ref}}}{Z+2Z_{0,\text{ref}}} & \dfrac{Z}{Z+2Z_{0,\text{ref}}} \end{bmatrix}, \tag{3.181}$$

where

$$Z = 1/(j\omega C). \tag{3.182}$$

3.27.2 Interdigital Capacitor

The capacitance of the interdigital capacitor is given by (Gupta, Garg and Chadha, 1981)

$$C = l(\varepsilon_r + 1)[(N-3)A_1 + A_2]\ (\text{pF}), \tag{3.183}$$

where N is the total number of fingers,
$A_1 = 0.225\text{pF/inch}$, $A_2 = 0.252\text{pF/inch}$ and l is the length in inch.
Hence the S-parameters of the interdigital capacitor in two-port connection are given by

$$[S] = \begin{bmatrix} \dfrac{Z}{Z+2Z_{0,\text{ref}}} & \dfrac{2Z_{0,\text{ref}}}{Z+2Z_{0,\text{ref}}} \\ \dfrac{2Z_{0,\text{ref}}}{Z+2Z_{0,\text{ref}}} & \dfrac{Z}{Z+2Z_{0,\text{ref}}} \end{bmatrix}, \tag{3.184}$$

where

$$Z = 1/(j\omega C). \tag{3.185}$$

3.28 LUMPED ELEMENTS: RESISTOR

Figure 3.28 shows the display of the Lumped Elements: Resistor device builder. The resistance of the resistor is given by

$$R = \begin{cases} R_s \dfrac{L_{\text{total}}}{2(W+t)} & \text{if } 2\delta_s < t \\ \dfrac{L_{\text{total}}}{\sigma W t} & \text{otherwise} \end{cases}, \tag{3.186}$$

where L_{total} is the total meander line length, R_s the surface resistance of the coil and δ_s the skin depth.

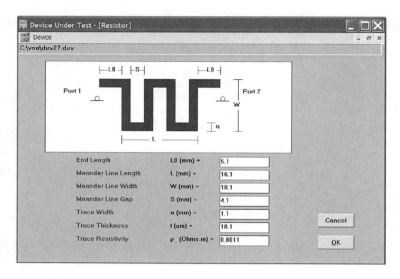

Figure 3.28   Lumped Elements: Resistor

3.29   ACTIVE DEVICES

Figure 3.29(a)–(d) shows the Active Devices device builders. The active device can be defined as a voltage-controlled voltage source, voltage-controlled current source, current-controlled voltage source, current-controlled current source. The S-parameters of these devices are, respectively (Dobrowolski, 1991)

$$[S]_{V \to V} = \begin{bmatrix} 1 & 0 \\ 2k & -1 \end{bmatrix}, \quad [S]_{V \to I} = \begin{bmatrix} 1 & 0 \\ -2Z_0 g & 1 \end{bmatrix}, \quad [S]_{I \to V} = \begin{bmatrix} -1 & 0 \\ 2R/Z_0 & -1 \end{bmatrix},$$

$$[S]_{I \to I} = \begin{bmatrix} -1 & 0 \\ -2b & -1 \end{bmatrix}.$$

$$(3.187)$$

3.30   ANTENNAS: DIPOLE ANTENNA

Figure 3.30 shows the display of the Antennas: Dipole Antenna device builder. In terms of circuit, the antenna can be modelled as load impedance given by (Jordan, 1950; Kraus, 1989)

$$R_{in}(b) = \frac{30}{\sin^2(b/2)} [2(1 + \cos b) S_i(b) - \cos b S_1(2b) - 2 \sin b S_i(b) + \sin b S_i(2b)]$$

$$= \frac{30}{\sin^2(b/2)} A(b),$$

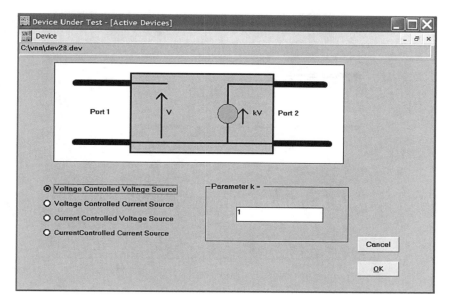

(a)

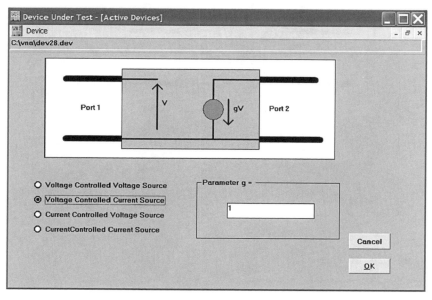

(b)

Figure 3.29 Active Devices: (a) voltage-controlled voltage source, (b) voltage-controlled current source, (c) current-controlled voltage source and (d) current-controlled current source

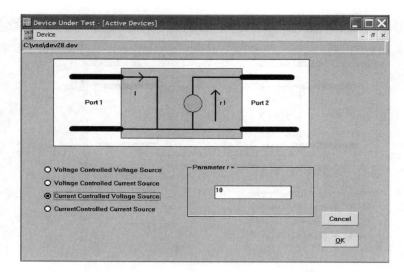

(c)

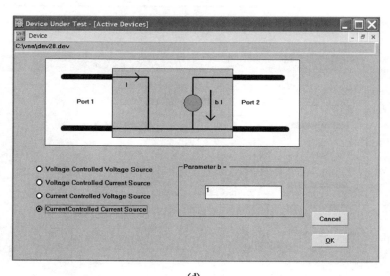

(d)

Figure 3.29 (Continued)

$$X_{in}(b) = \frac{-30}{\sin^2(b/2)} \left\{ (\sin b) \ln (H/\beta a^2) + \sin b \left[-\gamma + 2C_i(b) - C_i(2b) \right] \right.$$

$$\left. - \cos b \left[2S_i(b) - S_i(2b) \right] - 2S_i(b) \right\} = \frac{-30}{\sin^2(b/2)} \left[(\sin b) \ln \left(\frac{L^2}{2ba^2} \right) + B(b) \right],$$

$$(3.188)$$

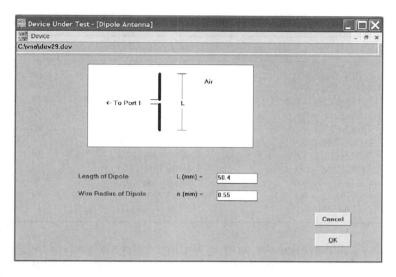

Figure 3.30   Antennas: Dipole Antenna

where

$$A(b) = 2(1+\cos b)S_i(b) - \cos bS_1(2b) - 2\sin bS_i(b) + \sin bS_i(2b),$$

$$B(b) = \sin b[-\gamma + 2C_i(b) - C_i(2b)] - \cos b[2S_i(b) - S_i(2b)] - 2S_i(b),$$

$$b = 2\beta H, \quad H = L/2, \quad \beta = \frac{2\pi}{\lambda} = \frac{2\pi f}{c},$$

$$S_1(x) = \int_0^x \frac{1-\cos v}{v}dv = \left(\frac{x^2}{2\cdot 2!} - \frac{x^4}{4\cdot 4!} + \frac{x^6}{6\cdot 6!} - \cdots + \cdots\right) = \frac{x^2}{4} \quad \text{for } x < 0.6,$$

$$S_1(x) \approx 1.957 + 0.08b - 0.1\sin(b) \quad \text{for } x > 6,$$

$$S_i(x) = \int_0^x \frac{\sin v}{v}dv = \left(x - \frac{x^3}{3!3} + \frac{x^5}{5!5} - \cdots\right) = \begin{cases} x & x < 0.5 \\ \dfrac{\pi}{2} - \dfrac{\cos x}{x} & x \gg 1 \end{cases},$$

$$C_i(x) = \ln x + 0.5772157 - S_1(x) = \int_0^x \frac{\cos v}{v}dv = \begin{cases} 0.577 + \ln x & \text{for } x < 0.2 \\ \dfrac{\sin x}{x} & \text{for } x \gg 1 \end{cases}.$$

$$(3.189)$$

Table 3.3 $A(b)$ and $B(b)$ expansion coefficients

A (b)	B (b)	Term
$-1.7297360e-05$	$7.9728378e-07$	b^{10}
$7.8565342e-04$	$-2.6679835e-05$	b^9
$-1.4319319e-02$	$4.3777030e-04$	b^8
$1.3680252e-01$	$-4.8503360e-03$	b^7
$-7.3924374e-01$	$3.6901743e-02$	b^6
$2.2557770e+00$	$-1.6801097e-01$	b^5
$-3.7426332e+00$	$3.8081304e-01$	b^4
$3.8275498e+00$	$-5.4009365e-01$	b^3
$-4.8368367e+00$	$2.1550133e+00$	b^2
$5.6086056e+00$	$-4.9904193e+00$	b^1
$-2.8144596e-01$	$-2.0374065e-01$	b^0

The calculation using the series expansion is very slow except for $b < 0.6$ or $b > 6$. For $0.6 < b < 6$, $A(b)$ and $B(b)$ are obtained by curve-fitting. The coefficients for b^0 to b^{10} are given in the Table 3.3 for both $A(b)$ and $B(b)$.

The S-parameter of the dipole antenna can then be obtained using the equations for a one-port terminated load described in Section 3.3, i.e.

$$S_{11} = \frac{Z_L - Z_{0,\text{ref}}}{Z_L + Z_{0,\text{ref}}} \tag{3.190}$$

with

$$Z_L = Z_{\text{in}} = R_{\text{in}} + jX_{\text{in}}. \tag{3.191}$$

3.31   ANTENNAS: RESONANT ANTENNA

Figure 3.31 shows the display of the Antennas: Resonant Antenna device builder. This model applies to a resonance type of antenna near the resonance frequency. The input admittance of the resonant antenna is defined as

$$Y_{\text{in}} == \frac{1}{R} + j\left(\omega C - \frac{1}{\omega L}\right) = \frac{1}{R} + jQ_0 R\left(f/f_0 - f_0/f\right), \tag{3.192}$$

where Q_0 is the unloaded Q-factor and f_0 the resonance frequency.

The S-parameter of the resonant antenna is then given by

$$S_{11} = \frac{Z_L - Z_{0,\text{ref}}}{Z_L + Z_{0,\text{ref}}} \tag{3.193}$$

with

$$Z_L = Y_{\text{in}}^{-1}. \tag{3.194}$$

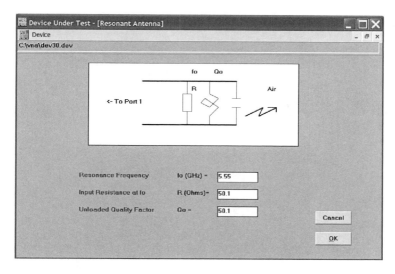

Figure 3.31   Antennas: Resonant Antenna

3.32   ANTENNAS: TRANSMISSION BETWEEN DIPOLE ANTENNAS

Figure 3.32 shows the display of the Antennas: Transmission between Dipole Antennas device builder. The reflection coefficients are given by

$$S_{11} = S_{22} = \frac{Z_{in} - Z_{0,ref}}{Z_{in} + Z_{0,ref}}, \tag{3.195}$$

where $Z_{in} = R_{in} + jX_{in}$ are calculated using the formulas in Section 3.30. Using the Radar Range Equation, the transmission coefficients can be obtained to be (Bryant, 1988)

$$S_{21} = S_{12} = |S_{21}| e^{-j(2\pi f/c)r}, \tag{3.196}$$

where

$$|S_{21}| = \left(1 - |S_{11}|^2\right)^{1/2} \left(1 - |S_{22}|^2\right)^{1/2} 1.64 \left(\frac{c}{4\pi f r}\right). \tag{3.197}$$

The gain of the dipole antenna is taken to be 1.64.

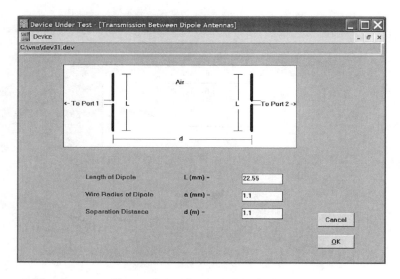

Figure 3.32 Antennas: Transmission between Dipole Antennas

3.33 ANTENNAS: TRANSMISSION BETWEEN RESONANT ANTENNAS

Figure 3.33 shows the display of the Antennas: Transmission between Resonant Antennas device builder. The reflection coefficients are given by

$$S_{11} = S_{22} = \frac{1/Y_{\text{in}} - Z_{0,\text{ref}}}{1/Y_{\text{in}} - Z_{0,\text{ref}}} = \frac{1 - Z_{0,\text{ref}}Y_{\text{in}}}{1 + Z_{0,\text{ref}}Y_{\text{in}}} \qquad (3.198)$$

where $Y_{\text{in}} = (1/R)\left[1 + jQ_0\left(f/f_0 - f_0/f\right)\right]$ as defined in Section 3.32. The transmission coefficients are given by

$$S_{21} = |S_{21}| \, e^{-j(2\pi f/c)r}, \qquad (3.199)$$

where

$$|S_{21}| = \left(1 - |S_{11}|^2\right)^{1/2} \left(1 - |S_{22}|^2\right)^{1/2} \sqrt{G_t G_r} \left(\frac{c}{4\pi f r}\right) = \left(1 - |S_{11}|^2\right) G \left(\frac{c}{4\pi f r}\right). \qquad (3.200)$$

The gain is taken to be G for both antennas.

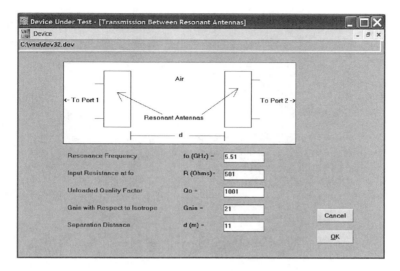

Figure 3.33   Antennas: Transmission between Resonant Antennas

3.34   USER-DEFINED *S*-PARAMETERS: ONE-PORT DEVICE

Figure 3.34 shows the display of the User-Defined *S*-Parameters: One-Port Device builder. The user-defined S_{11} parameters at different frequencies in the range between f_1 and f_M can be inputted, where the value of M is 51. These may include the measured S_{11} parameters or those of active devices. The Software VNA calculates the *S*-parameters of the device at different frequencies using the linear interpolation

$$S_{11,i}(f) = S_{11,i+1}\frac{f-f_i}{f_{i+1}-f_i} + S_{11,i}\frac{f_{i+1}-f}{f_{i+1}-f_i} \tag{3.201}$$

so that

$$S_{11}(f) = \begin{cases} \sum_{i=1}^{M-1} S_{11,i}(f)\, U(f_{i+1},f_i) & \text{if } f_1 \leq f \leq f_M \\ S_{11,1} & \text{if } f \leq f_1 \\ S_{11,M} & \text{if } f > f_M \end{cases}, \tag{3.202}$$

where

$$U(f_{i+1},f_i) = \begin{cases} 1 & \text{if } f_i < f \leq f_{i+1} \\ 0 \end{cases}. \tag{3.203}$$

In addition to the manual input, the one-port *S*-parameter saved using the SaveData function of the Software VNA for a simulated network or

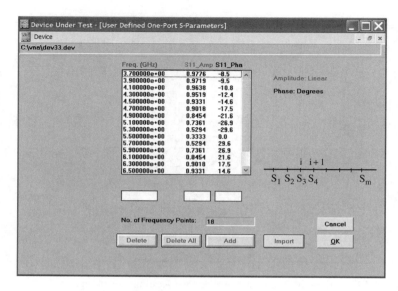

Figure 3.34   User-Defined S-Parameters: One-Port Device

circuit can be imported into the device builder to form a new device by clicking on the Import button. In this case, only 51 frequency points are selected from the saved data from the first point to the last point with equal frequency spacing.

3.35   USER-DEFINED S-PARAMETERS: TWO-PORT DEVICE

Figure 3.35 shows the display of the User-Defined S-Parameters: Two-Port Device builder. The user can define all four S-parameters at different frequencies in the range between f_1 and f_M, where the value of M is 51. These may include the measured S-parameters or those of active devices. The Software VNA calculates the S-parameters of the device at different frequencies using the linear interpolation

$$S_i(f) = S_{i+1}\frac{f-f_i}{f_{i+1}-f_i} + S_i\frac{f_{i+1}-f}{f_{i+1}-f_i} \tag{3.204}$$

so that

$$S(f) = \begin{cases} \sum_{i=1}^{M-1} S_i(f)\,U(f_{i+1},f_i) & \text{if } f_1 \le f \le f_M \\ S_1 & \text{if } f \le f_1 \\ S_M & \text{if } f > f_M \end{cases}, \tag{3.205}$$

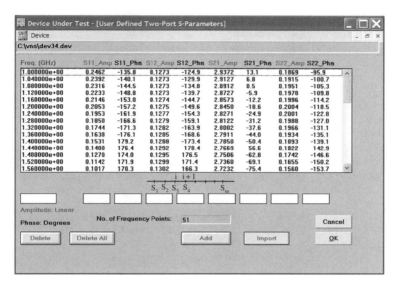

Figure 3.35   User-Defined S-Parameters: Two-Port Device

where

$$U\left(f_{i+1}, f_i\right) = \begin{cases} 1 & \text{if } f_i < f \le f_{i+1} \\ 0 \end{cases}. \tag{3.206}$$

and S represents each of the four S-parameters.

In addition to the manual input, the two-port S-parameters saved using the SaveData function of the Software VNA for a simulated network or circuit can be imported into the device builder to form a new device by clicking on the Import button. In this case, only 51 frequency points are selected from the saved data from the first point to the last point with equal frequency spacing.

REFERENCES

Alexopolous, N.G. and Wu, S. (1994) 'Frequency-independent equivalent circuit model for microstrip open-end and gap discontinuities'. *IEEE Transactions on Microwave Theory and Techniques*, **MTT-42** (7), 1268–72.

Bahl, I.J. (1989) Transmission lines, in *Handbook of Microwave and Optical Components – Vol. 1: Microwave Passive and Antenna Components* (ed. K. Chang), John Wiley & Sons, Ltd, New York, pp. 1–59.

Bryant, G.H. (1988) *Principles of Microwave Measurements*, IEE, London.

Dobrowolski, J.A. (1991) *Introduction to Computer Methods for Microwave Circuit Analysis and Design*, Artech House, Boston.

Gupta, K.C., Garg, R. and Chadha, R. (1981) *Computer Aided Design of Microwave Circuits*, Artech House, Norwood.

Gupta, K.C., Garg, R., Bahl, I. and Bhartia, P. (1996) *Microstrip Lines and Slotlines*, 2nd edn, Artech House, Norwood.

Howe, H., Jr (1974) *Stripline Circuit Design*, Artech House, Dedham, MA.

Jordan, E.C. (1950)*Electromagnetic Waves and Radiating Systems*, Prentice-Hall, Englewood Cliffs, NJ.

Kraus, J.D. (1989) *Antennas*, McGraw-Hill, New York.

Marcuvitz, N. (1951) *Waveguide Handbook*, McGraw-Hill, New York.

Medhurst, R.G. (1947) 'H.F. resistance and self-capacitance of single-layer solenoids'. *Wireless Engineer*, 80–92.

Misra, D. et al. (1990) 'Noninvasive electrical characterisation of materials at microwave frequencies using an open-ended coaxial line: test of an improved calibration technique'. *IEEE Transactions on Microwave Theory and Techniques*, MTT-38, 8–14.

Napoli, L.S. and Hughes, J.J. (1970) 'Characteristics of coupled microstrip lines'. *RCA Review*, 479–98.

Pozar, D.M. (1990) *Microwave Engineering*, Addison-Wesley, New York.

Wadell, B.C. (1991) *Transmission Line Design Handbook*, Artech House, Norwood, 1991.

4

Design of Microwave Circuits

ABSTRACT

The Circuit Simulator in the Software VNA can be used to simulate microwave circuits assembled with the devices built using the device builders in the software package. In this chapter, the design principles of a number of commonly used microwave circuits impedance matching, impedance transformers, microwave resonators, power dividers, couplers, hybrid rings, phase shifters, filters and amplifier design are described.

KEYWORDS

Microwave circuits, Impedance matching, Impedance transformers, Microwave resonators, Power dividers, Couplers, Hybrid rings, Phase shifters, Filters, Amplifier design

The Circuit Simulator in the Software VNA can be used to simulate microwave circuits assembled with the devices built using the device builders described in Chapter 3. In this chapter, the design principles of a number of commonly used microwave circuits are described. Examples of circuit simulation using the Circuit Simulator will be presented in Chapter 5.

4.1 IMPEDANCE MATCHING

4.1.1 Impedance Matching Using a Discrete Element

The use of a quarter-wave transformer and a discrete element is the simplest technique for impedance matching. The diagram of this impedance matching technique is shown in Figure 4.1. The load impedance is denoted as $Z_L = R_L + jX_L$ and admittance $Y_L = 1/Z_L = G_L + jB_L$. If the load

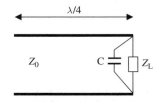

Figure 4.1   Impedance matching using a discrete element

impedance is inductive, a discrete capacitor of capacitance C is connected in parallel to the load impedance so that the overall impedance is resistive at the required matching frequency f_0. This leads to

$$C = -\frac{B}{2\pi f_0} \qquad (4.1)$$

and the resultant resistance of $1/G_L$. A quarter-wave transformer is then used to transform the resultant resistance to the source impedance of $Z_s = 50\,\Omega$. The characteristic impedance of the quarter-wave transformer is required to be

$$Z_0 = \sqrt{50/G_L}. \qquad (4.2)$$

If the load impedance is capacitive, an inductor can be used to replace the capacitor. The load impedance can then be matched to the source impedance in a similar way. A disadvantage of this matching technique is that it requires the design and fabrication of different quarter-wave transformers if the load impedance is changed.

4.1.2   Single Stub Matching

The principle of single stub matching is to transform the load impedance $Z_L = R_L + jX_L$ using a section of transmission line of length L_1 and standard characteristic impedance, e.g. $Z_0 = 50\,\Omega$ to an admittance of $Y_{in1} = 0.02 + jB$ (Collin, 1966). An open- or short-circuited stub of the same characteristic impedance but length L_2 is then connected in parallel to the equivalent admittance with an input admittance of $Y_{in2} = -jB$ to cancel the reactive part of Y_{in1}. This results in the impedance matching to the source impedance of $Z_s = 50\,\Omega$. The diagram of the single stub matching is shown in Figure 4.2. The required lengths L_1 and L_2 can be calculated using

$$\frac{L_1}{\lambda} = \begin{cases} \frac{1}{2\pi}\tan^{-1} t & \text{if } t \geq 0 \\ \frac{1}{2\pi}\left(\pi + \tan^{-1} t\right) & \text{if } t < 0 \end{cases} \qquad (4.3)$$

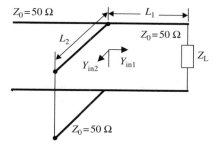

Figure 4.2   Single stub matching

and

$$\frac{L_2}{\lambda} = \frac{+1}{2\pi} \tan^{-1}\left(\frac{1}{Z_0 B}\right) \tag{4.4}$$

for a short-circuited stub where

$$t = \tan(\beta d) = \frac{X_L \pm \sqrt{R_L\left[(Z_0 - R_L)^2 + X_L^2\right]/Z_0}}{R_L - Z_0} \qquad (R_L \neq Z_0) \tag{4.5}$$

and

$$B = \frac{R_L^2 t - (Z_0 - X_L t)(X_L + Z_0 t)}{Z_0\left[R_L^2 + (X_L + Z_0 t)^2\right]}. \tag{4.6}$$

There are two possible solutions of L_1, and consequently L_2. The length L_2 may also be different if an open-circuited stub is used. The disadvantage of the single stub matching technique is that it requires different transmission line lengths when the load impedance is changed.

4.1.3 Double Stub Matching

The double stub matching technique may overcome the disadvantage of the single stub matching (Collin, 1966; Pozar, 1990). In this technique, two stubs are used. The first stub is connected in parallel with the load impedance $Z_L = R_L + jX_L$ or admittance $Y_L = 1/Z_L = G_L + jB_L$. A fixed length of transmission line, d, is used between the load and the second stub. The diagram of this matching technique is shown in Figure 4.3.

The length of the transmission line between two stubs is chosen so that

$$0 \leq G_L \leq Y_0 \frac{1 + t^2}{2t^2} = \frac{Y_0}{\sin^2 \beta d}, \tag{4.7}$$

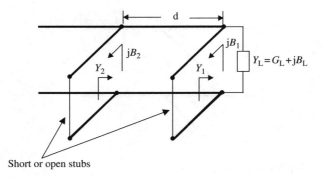

Figure 4.3   Double stub matching

where $t = \tan{(\beta d)}$ and Y_0 is the characteristic admittance of the transmission line. In practice, d is usually chosen to be $\lambda/8$ or $3\lambda/8$. The length of the first stub is required to be

$$\frac{l_{s1}}{\lambda} = \begin{cases} \frac{-1}{2\pi}\tan^{-1}\left(\frac{Y_0}{B_1}\right) & \text{for short-circuited stub} \\ \frac{1}{2\pi}\tan^{-1}\left(\frac{B_1}{Y_0}\right) & \text{for open-circuited stub} \end{cases} \tag{4.8}$$

in which

$$B_1 = -B_L + \frac{Y_0 \pm \sqrt{(1+t^2)G_L Y_0 - G_L^2 t^2}}{t}. \tag{4.9}$$

Similarly the length of the second stub is required to be

$$\frac{l_{s2}}{\lambda} = \begin{cases} \frac{-1}{2\pi}\tan^{-1}\left(\frac{Y_0}{B_2}\right) & \text{for short-circuited stub} \\ \frac{1}{2\pi}\tan^{-1}\left(\frac{B_2}{Y_0}\right) & \text{for open-circuited stub} \end{cases}, \tag{4.10}$$

where

$$B_2 = \frac{\pm Y_0\sqrt{Y_0 G_L(1+t^2) - G_L^2 t^2} + G_L Y_0}{G_L t}. \tag{4.11}$$

The input admittance after the first stub is

$$Y_1 = G_L + j(B_L + B_1) \tag{4.12}$$

and that before the second stub is

$$Y_2 = Y_0 \frac{G_L + j(B_L + B_1 + Y_0 t)}{Y_0 + jt(G_L + jB_L + jB_1)} = Y_0 - jB_2. \tag{4.13}$$

The input impedance after the second stub is therefore Z_0, which should be equal to the source impedance Z_s.

4.2 IMPEDANCE TRANSFORMERS

4.2.1 Quarter-Wave Transformer

The quarter-wave transformer, as shown in Figure 4.4, is usually used to match a resistive load $Z_L = R_L$ to a source impedance Z_s. The length of the transformer is $\lambda/4$ at the operating frequency f_0. The characteristic impedance of the transformer is

$$Z_0 = \sqrt{Z_s Z_L}. \tag{4.14}$$

In general, the input impedance of the transformer is (Collin, 1966; Pozar, 1990)

$$Z_{in} = Z_0 \frac{Z_L + jZ_0 t}{Z_0 + jZ_L t}, \tag{4.15}$$

where $t = \tan(\beta l) = \tan\theta$ and $\beta = 2\pi/\lambda$ with $\beta l = \theta = \pi/2$ at f_0. This gives a reflection coefficient of

$$\Gamma = \frac{Z_L - Z_s}{Z_L + Z_s + j2t\sqrt{Z_s Z_L}}. \tag{4.16}$$

The fractional bandwidth measured at the level $|\Gamma| = \Gamma_m$ can be determined to be

$$\frac{\Delta f}{f_0} = \frac{2(f_0 - f_m)}{f_0} = 2 - \frac{4}{\pi}\cos^{-1}\left[\frac{\Gamma_m}{\sqrt{1 - \Gamma_m^2}} \frac{2\sqrt{Z_0 Z_L}}{|Z_L - Z_0|}\right]. \tag{4.17}$$

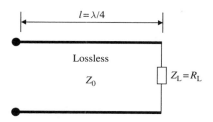

Figure 4.4 Quarter-wave transformer

4.2.2   Chebyshev Multisection Matching Transformer

The Chebyshev multisection matching transformer, as shown in Figure 4.5, is also used to match a resistive load $Z_L = R_L$ to a source impedance Z_s (Collin, 1966; Pozar, 1990). Each transmission line section is $\lambda/4$ long at the operating frequency f_0. The characteristic impedance of each line section is to be determined.

In general, the reflection coefficient of the terminated multisection transmission line system is given by

$$\Gamma(\theta) = \begin{cases} 2e^{-jN\theta} \left[\Gamma_0 \cos N\theta + \Gamma_1 \cos(N-2)\theta + \cdots + \Gamma_n \cos(N-2n)\theta \right. \\ \left. \qquad + \cdots + \tfrac{1}{2}\Gamma_{N/2}\right] \qquad\qquad\qquad\qquad \text{for even } N \\ 2e^{-jN\theta} \left[\Gamma_0 \cos N\theta + \Gamma_1 \cos(N-2)\theta + \cdots + \Gamma_n \cos(N-2n)\theta \right. \\ \left. \qquad + \cdots + \Gamma_{(N-1)/2} \cos\theta\right] \qquad\qquad \text{for odd } N \end{cases}$$

(4.18)

where

$$\Gamma_0 = \frac{Z_1 - Z_0}{Z_1 + Z_0}, \quad \Gamma_n = \frac{Z_{n+1} - Z_n}{Z_{n+1} + Z_n}, \quad \Gamma_N = \frac{Z_L - Z_N}{Z_L + Z_N} \qquad (4.19)$$

and $Z_0 = Z_s$ is the source impedance, Z_n the characteristic impedance of the nth section. With the assumption that the transformer is symmetrical, the following equalities can be established:

$$\Gamma_0 = \Gamma_N, \quad \Gamma_1 = \Gamma_{N-1}, \quad \Gamma_2 = \Gamma_{N-2} \quad \text{and so on.} \qquad (4.20)$$

The characteristic impedance of each transmission line section can be determined when Γ_n values are assigned. Using the Chebyshev polynomials which can be expressed as

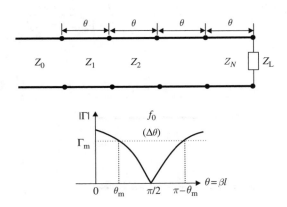

Figure 4.5   Chebyshev multisection matching transformer

$$T_N(x) = \cos\left(N\cos^{-1}x\right) \text{ for } |x| < 1 \text{ and } T_N(x)$$
$$= \cosh\left(N\cosh^{-1}x\right) \text{ for } |x| > 1 \tag{4.21}$$

with $|T_N(x = \pm 1)| = 1$, the reflection coefficient can be written as

$$\Gamma(\theta) = \Gamma_m e^{-jN\theta} T_N(\sec\theta_m \cos\theta), \tag{4.22}$$

where N is the number of sections, θ a frequency-dependent quantity which is related to the electric length of the line section, Γ_m the maximum allowable reflection coefficient in the matched band which occurs at θ_m and $\pi - \theta_m$ where

$$\sec\theta_m = \cosh\left[\frac{1}{N}\cosh^{-1}\left(\frac{1}{\Gamma_m}\left|\frac{Z_L - Z_0}{Z_L + Z_0}\right|\right)\right]. \tag{4.23}$$

At the central operating frequency f_0, $\theta = \pi/2$ and the length of the line section is $\lambda/4$. The fractional bandwidth of the Chebyshev multisection matching transformer is given by

$$\frac{\Delta f}{f_0} = 2 - \frac{4\theta_m}{\pi}. \tag{4.24}$$

In the case that $N = 2$,

$$\Gamma(\theta) = \Gamma_m e^{-j2\theta} T_2(\sec\theta_m \cos\theta) = e^{-j2\theta}\Gamma_m\left[\sec^2\theta_m(1 + \cos 2\theta) - 1\right]$$
$$= e^{-j2\theta}\Gamma_m\left[\sec^2\theta_m \cos 2\theta + (\sec^2\theta_m - 1)\right]$$
$$= 2e^{-j2\theta}\left[\Gamma_0 \cos 2\theta + \frac{\Gamma_1}{2}\right] \tag{4.25}$$

and

$$\sec\theta_m = \cosh\left[\frac{1}{2}\cosh^{-1}\left(\frac{1}{\Gamma_m}\left|\frac{Z_L - Z_0}{Z_L + Z_0}\right|\right)\right]. \tag{4.26}$$

Hence

$$\Gamma_0 = \Gamma_2 = \frac{1}{2}\Gamma_m \sec^2\theta_m \text{ and } \Gamma_1 = \Gamma_m(\sec^2\theta_m - 1) \tag{4.27}$$

from which Z_1 and Z_2 can be calculated.

In the case that $N = 3$,

$$\Gamma(\theta) = \Gamma_m e^{-j3\theta} T_3(\sec\theta_m \cos\theta)$$
$$= e^{-j3\theta}\Gamma_m\left[\sec^3\theta_m \cos 3\theta + 3\sec^3\theta_m \cos\theta - 3\sec\theta_m \cos\theta\right]$$
$$= e^{-j3\theta}\Gamma_m\left[\sec^3\theta_m \cos 3\theta + 3\sec\theta_m(\sec^2\theta_m - 1)\cos\theta\right]$$
$$= 2e^{-j3\theta}[\Gamma_0 \cos 3\theta + \Gamma_1 \cos\theta] \tag{4.28}$$

and

$$\sec\theta_m = \cosh\left[\frac{1}{3}\cosh^{-1}\left(\frac{1}{\Gamma_m}\left|\frac{Z_L - Z_0}{Z_L + Z_0}\right|\right)\right]. \qquad (4.29)$$

Hence

$$\Gamma_0 = \Gamma_3 = \frac{1}{2}\Gamma_m \sec^3\theta_m \quad\text{and}\quad \Gamma_1 = \Gamma_2 = \frac{1}{2}\Gamma_m \cdot 3\sec\theta_m(\sec^2\theta_m - 1) \qquad (4.30)$$

from which Z_1, Z_2 and Z_3 can be calculated.

In the case that $N = 4$,

$$\begin{aligned}
\Gamma(\theta) &= \Gamma_m T_4(\sec\theta_m \cos\theta) = 2\left[\Gamma_0 \cos 4\theta + \Gamma_1 \cos(2\theta) + \frac{1}{2}\Gamma_2\right] \\
&= \Gamma_m[\sec^4\theta_m \cos 4\theta + 4\sec^4\theta_m \cos 2\theta + 3\sec^4\theta_m - 4\sec^2\theta_m \cos 2\theta \\
&\qquad + 1 - 4\sec^2\theta_m] \\
&= \Gamma_m[\sec^4\theta_m \cos 4\theta + 4\sec^2\theta_m(\sec^2\theta_m - 1)\cos 2\theta + 1 + 3\sec^4\theta_m \\
&\qquad - 4\sec^2\theta_m] \qquad (4.31)
\end{aligned}$$

so that

$$\Gamma_0 = \Gamma_4 = \frac{1}{2}\Gamma_m \sec^4\theta_m, \quad \Gamma_1 = \Gamma_3 = 2\Gamma_m \sec^2\theta_m(\sec^2\theta_m - 1) \text{ and}$$
$$\Gamma_2 = \Gamma_m(1 + 3\sec^4\theta_m - 4\sec^2\theta_m). \qquad (4.32)$$

The characteristic impedance of each line section, Z_1, Z_2, Z_3 and Z_4, can then be calculated.

4.2.3 Corporate Feeds

Figure 4.6(a) and (b) shows the two basic configurations of corporate feeds which are commonly used in antenna arrays. The antenna array can be matched to a $50\,\Omega$ feed line by the successive application of the basic corporate feed.

In the first configuration, the $50\,\Omega$ arms and feed line can be of any length. The impedance transformer is, however, $\lambda/4$ long. It transforms the resultant impedance of two $50\,\Omega$ impedances connected in parallel to the impedance of the feed line, i.e. $50\,\Omega$, so that the transformer has a characteristic impedance of

$$Z_0 = \sqrt{25 \times 50} = 35.4\,\Omega. \qquad (4.33)$$

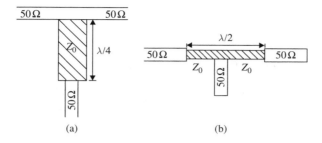

Figure 4.6 Corporate feed: (a) configuration 1 and (b) configuration 2

In the second configuration, the $50\,\Omega$ arms and feed line can be of any length. It consists of two quarter-wave impedance transformers. Each transformer transforms the $50\,\Omega$ impedance to $100\,\Omega$. Two $100\,\Omega$ impedances are then connected in parallel to give a resultant impedance of $50\,\Omega$, which is then matched to the impedance of the feed line. The total length of the impedance transformers is therefore $\lambda/2$, and the characteristic impedance of the transformers is

$$Z_0 = \sqrt{50 \times 100} = 70.7\,\Omega. \tag{4.34}$$

4.3   MICROWAVE RESONATORS

4.3.1   One-Port Directly Connected RLC Resonant Circuits

The elements, R, L and C, connected in series and in parallel, as shown in Figure 4.7(a) and (b), are the two most fundamental resonant circuits. The resonance circuit is connected to the input port directly with $Z_s = Z_0$.

The resonance frequency of the RLC series resonant circuit shown in Figure 4.7(a) is (Wu and Davis, 1994; Ginzton, 1957)

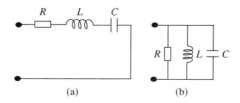

Figure 4.7 One-port directly coupled resonators: (a) R, L and C in series and (b) R, L and C in parallel

$$f_0 = \frac{1}{2\pi\sqrt{LC}} \qquad (4.35)$$

and unloaded and loaded Q-factors are, respectively,

$$Q_0 = \frac{2\pi f_0 L}{R} \quad \text{and} \quad Q_L = \frac{2\pi f_0 L}{R+Z_0} = \frac{Q_0}{1+\beta_1}, \qquad (4.36)$$

where β_1 is the coupling coefficient and

$$\beta_1 = \frac{Z_0}{R}. \qquad (4.37)$$

The resonance frequency of the RLC parallel resonant circuit shown in Figure 4.7(b) is

$$f_0 = \frac{1}{2\pi\sqrt{LC}} \qquad (4.38)$$

and unloaded and loaded Q-factors are, respectively,

$$Q_0 = \frac{R}{2\pi f_0 L} = 2\pi f_0 CR \quad \text{and} \quad Q_L = \frac{2\pi f_0 C}{R^{-1}+Z_0^{-1}} = \frac{Q_0}{1+(R/Z_0)} = \frac{Q_0}{1+\beta_1}, \qquad (4.39)$$

where

$$\beta_1 = \frac{R}{Z_0}. \qquad (4.40)$$

4.3.2   Two-Port Directly Connected RLC Resonant Circuits

The RLC series and parallel resonant circuits in two-port connections are shown in Figure 4.8(a) and (b), respectively, with $Z_s = Z_0$.

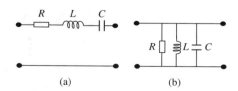

$$\text{(a)} \qquad \qquad \text{(b)}$$

Figure 4.8   Two-port directly coupled resonators: (a) R, L and C in series and (b) R, L, and C in parallel

The resonance frequency of the RLC series resonant circuit shown in Figure 4.8(a) is (Wu and Davis, 1994; Ginzton, 1957)

$$f_0 = \frac{1}{2\pi\sqrt{LC}} \tag{4.41}$$

and unloaded and loaded Q-factors are, respectively,

$$Q_0 = \frac{2\pi f_0 L}{R} \quad \text{and} \quad Q_L = \frac{2\pi f_0 L}{R+Z_0} = \frac{Q_0}{1+2\beta_1}, \tag{4.42}$$

where

$$\beta_1 = \frac{Z_0}{R}. \tag{4.43}$$

The resonance frequency of the RLC parallel resonant circuit shown in Figure 4.8(b) is

$$f_0 = \frac{1}{2\pi\sqrt{LC}} \tag{4.44}$$

and unloaded and loaded Q-factors are, respectively,

$$Q_0 = \frac{R}{2\pi f_0 L} = 2\pi f_0 CR \quad \text{and} \quad Q_L = \frac{2\pi f_0 C}{R^{-1}+2Z_0^{-1}} = \frac{Q_0}{1+(2R/Z_0)} = \frac{Q_0}{1+2\beta_1}, \tag{4.45}$$

where

$$\beta_1 = \frac{R}{Z_0}. \tag{4.46}$$

4.3.3 One-Port Coupled Resonators

The equivalent circuit of a microwave resonator is often represented by R, L and C elements connected in series, as shown in Figure 4.9. The coupling of the resonator to the input port with reference impedance Z_0 is represented by a transformer with the coupling ratio of $1:n_1$.

The resonance frequency and unloaded Q-factor are (Wu and Davis, 1994; Ginzton, 1957)

$$f_0 = \frac{1}{2\pi\sqrt{LC}} \tag{4.47}$$

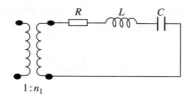

Figure 4.9   One-port coupled resonator

and

$$Q_0 = \frac{2\pi f_0 L}{R}, \tag{4.48}$$

respectively. The loaded Q-factor is

$$Q_L = \frac{Q_0}{1+\beta_1} \tag{4.49}$$

where

$$\beta_1 = \frac{n_1{}^2 Z_0}{R} \tag{4.50}$$

is the coupling coefficient. The S_{11} response of the one-port coupled resonator is

$$S_{11}(f) = \frac{S_{11}(f_0) + jQ_L\delta(f)}{1 + jQ_L\delta(f)}, \tag{4.51}$$

where

$$S_{11}(f_0) = \frac{1-\beta_1}{1+\beta_1} \quad \text{and} \quad \delta(f) = \left(\frac{f}{f_0} - \frac{f_0}{f}\right) = 2(f - f_0)/f_0. \tag{4.52}$$

The loaded and unloaded Q-factors can be measured from the amplitude and/or phase responses of S_{11} (Wu and Davis, 1994).

4.3.4   Two-Port Coupled Resonators

The equivalent circuit of a two-port coupled microwave resonator is shown in Figure 4.10. The couplings of the resonator to the input and output ports are represented by a transformer with the coupling ratio of $1:n_1$ and $n_2:1$, respectively. The port impedance of both ports is Z_0.

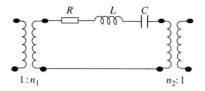

Figure 4.10   Two-port coupled resonator

The resonance frequency and unloaded Q-factor are (Wu and Davis, 1994; Ginzton, 1957)

$$f_0 = \frac{1}{2\pi\sqrt{LC}} \tag{4.53}$$

and

$$Q_0 = \frac{2\pi f_0 L}{R}, \tag{4.54}$$

respectively. The loaded Q-factor is

$$Q_L = \frac{Q_0}{1 + \beta_1 + \beta_2}, \tag{4.55}$$

where

$$\beta_1 = \frac{n_1^2 Z_0}{R} \tag{4.56}$$

is the coupling coefficient for the input port and

$$\beta_2 = \frac{n_2^2 Z_0}{R} \tag{4.57}$$

is the coupling coefficient for the output port. The S_{21} response of the one-port coupled resonator is

$$S_{21}(f) = \frac{S_{21}(f_0)}{1 + jQ_L \delta(f)}, \tag{4.58}$$

where

$$S_{21}(f_0) = \frac{2(\beta_1 \beta_2)^{1/2}}{1 + \beta_1 + \beta_2} \quad \text{and} \quad \delta(f) = \left(\frac{f}{f_0} - \frac{f_0}{f}\right) = 2(f - f_0)/f_0. \tag{4.59}$$

For a symmetrically coupled resonator,

$$\beta_1 = \beta_2 = \beta \tag{4.60}$$

so that

$$S_{21}(f_0) = \frac{2\beta}{1+2\beta}.$$ (4.61)

The loaded and unloaded Q-factors can be measured from the amplitude or phase responses of S_{21} (Wu and Davis, 1994).

4.3.5   Transmission Line Resonators

A short- or open-circuited transmission line may resonate at frequencies when it is $n(\lambda/4)$ long where $n = 1, 2, \ldots$. A transmission line with short- or open- circuits at both ends may resonate at frequencies when it is $n(\lambda/2)$ long (Collin, 1966; Pozar, 1990).

The unloaded Q-factor of the transmission line is

$$Q_0 = \beta/2\alpha,$$ (4.62)

where β is the phase constant and α the attenuation constant of the transmission line at the resonance frequency.

The transmission line resonator can be coupled using either one-port configuration discussed in Section 4.3.3 or two-port configuration in Section 4.3.4. A two-port coupled microstrip line resonator is shown in Figure 4.11. The couplings are realised by two capacitive gaps.

4.3.6   Coupled Line Resonators

Since both even and odd modes can be excited on a coupled line simultaneously, two adjacent resonant modes may be observed: one corresponds to the even mode resonance and the other to the odd mode resonance.

For a two-port coupled coupled line resonator as shown in Figure 4.12, the first resonance pair will occur when the length of the line is $(\lambda_e/2)$ or $(\lambda_o/2)$, which gives rise to the resonance frequencies for the even and odd modes.

Figure 4.11   Two-port coupled microstrip line resonator

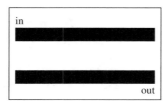

Figure 4.12 Two-port coupled coupled line resonator

The unloaded Q-factors for the even and odd modes of the coupled line resonator are, respectively,

$$Q_{0,e} = \beta_e / 2\alpha_e \qquad (4.63)$$

and

$$Q_{0,o} = \beta_o / 2\alpha_o. \qquad (4.64)$$

4.4 POWER DIVIDERS

Power dividers are used in microwave circuits to control the power flow with impedance matching at all ports. They are made of transmission lines with specific lengths and characteristic impedances. A number of microwave power dividers are described below.

4.4.1 The 3 dB Wilkinson Power Divider

Figure 4.13 shows the design of a 3 dB Wilkinson power divider (Pozar, 1990; Bhat and Koul, 1989). The input power from Port 1 is split equally between two output ports, Port 2 and Port 3. Ports 2 and 3 are isolated. The S-parameters of the 3 dB power divider are

$$[S] = \begin{bmatrix} 0 & -j\frac{1}{\sqrt{2}} & -j\frac{1}{\sqrt{2}} \\ -j\frac{1}{\sqrt{2}} & 0 & 0 \\ -j\frac{1}{\sqrt{2}} & 0 & 0 \end{bmatrix}. \qquad (4.65)$$

The port impedance for all three ports is considered to be Z_0. The length and characteristic impedance of both arms are $\lambda/4$ and $\sqrt{2}Z_0$, respectively. The resistance of the isolation resistor connecting the two arms is $R = 2Z_0$.

The bandwidth of the 3 dB Wilkinson power divider is BW $= 1.44 : 1$ measured at VSWR $= 1.22$. The isolation between Ports 2 and 3 is usually greater than 20 dB.

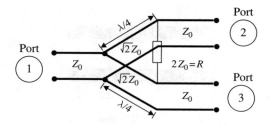

Figure 4.13 The 3 dB Wilkinson power divider

4.4.2   The Wilkinson Power Divider with Unequal Splits

Figure 4.14 shows the design of a Wilkinson power divider with unequal splits (Pozar, 1990; Bhat and Koul, 1989). The input power from Port 1 is split between two output ports, Port 2 and Port 3, with a ratio defined as

$$K^2 = \frac{P_3}{P_2}. \tag{4.66}$$

All transmission line sections in the divider are $\lambda/4$ in length, and the characteristic impedances of the line sections are

$$Z_2 = Z_0 \left[K \left(1 + K^2 \right) \right]^{1/2}, \quad Z_3 = Z_0 \left(\frac{1 + K^2}{K^3} \right)^{1/2}, \quad Z_4 = Z_0 \sqrt{K} \quad \text{and}$$

$$Z_5 = Z_0 \Big/ \sqrt{K}. \tag{4.67}$$

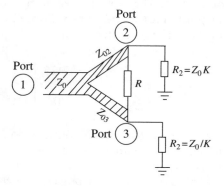

Figure 4.14 The 3 dB Wilkinson power divider with unequal splits in microstrip line layout

The resistance of the isolation resistor is

$$R = Z_0 \frac{1+K^2}{K}. \tag{4.68}$$

The bandwidth of the power divider is maximum for equal power split and becomes narrower as the ratio K increases.

4.4.3 Alternative Design of Power Divider with Unequal Splits

Figure 4.15 shows an alternative design of the power divider with unequal splits (Bhat and Koul, 1989; Parad and Moynihan , 1965). The input power from Port 1 is split between two output ports, Port 2 and Port 3, with a ratio

$$\frac{P_3}{P_2} = K^2. \tag{4.69}$$

All transmission line sections in the divider are $\lambda/4$ in length, and the characteristic impedances of the line sections are

$$Z_1 = Z_0 \left(\frac{K}{1+K^2} \right)^{1/4}, \quad Z_2 = Z_0 K^{3/4} \left(1+K^2 \right)^{1/4}, \quad Z_3 = \frac{Z_0 \left(1+K^2 \right)^{1/4}}{K^{4/5}},$$

$$Z_4 = Z_0 \sqrt{K} \quad \text{and} \quad Z_5 = Z_0 \big/ \sqrt{K}. \tag{4.70}$$

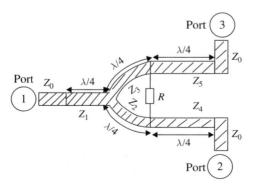

Figure 4.15 An alternative design of the 3 dB Wilkinson power divider with unequal splits in microstrip line layout

The resistance of the isolation resistor is

$$R = Z_0 \frac{1+K^2}{K}. \tag{4.71}$$

The bandwidth is also narrower when the ratio K increases.

4.4.4 Cohn's Cascaded Power Divider

Figure 4.16 shows the design of cascaded power divider of two sections proposed by Cohn (1968). The input power from Port 1 is split equally between two output ports, Port 2 and Port 3, with zero phase difference. All transmission sections in the divider are $\lambda/4$ in length. The characteristic impedances of the line sections for a frequency bandwidth ratio of $f_2/f_1 = 2$ defined at VSWR = 1.106 are given by

$$Z_1 = 1.2197 Z_0 \quad \text{and} \quad Z_2 = 1.6398 Z_0. \tag{4.72}$$

The resistance values of the isolation resistors are calculated by

$$R_2 = \frac{2 Z_1 Z_2}{\left[(Z_1 + Z_2)(Z_1 - Z_2 \cot^2 \phi) \right]^{1/2}}; \tag{4.73}$$

$$R_1 = \frac{2 R_2 (Z_1 + Z_2)}{R_2 (Z_1 + Z_2) - 2 Z_1}, \tag{4.74}$$

where

$$\phi = \frac{\pi}{2} \left[1 - \frac{1}{\sqrt{2}} \left(\frac{f_2 - f_1}{f_2 + f_1} \right) \right]. \tag{4.75}$$

Cascaded power dividers of multiple sections have also been formulated by Cohn (1968). Compared with single section power dividers, the cascaded power dividers have larger bandwidth and higher isolation between two output ports.

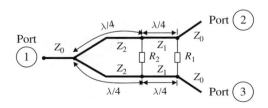

Figure 4.16 Cohn's cascaded power divider

4.5   COUPLERS

Couplers are also used in microwave circuits to control the power flow with impedance matching at all ports. They have multiple input and output ports. The diagram of a four-port coupler is shown in Figure 4.17. The parameters of the coupler with respect to the input from Port 1 is defined as follows (Bhat and Koul, 1989):

$$\text{Coupling } C(\text{dB}) = -10\log_{10}\frac{P_3}{P_2} = -20\log_{10}|S_{31}|$$

$$\text{Directivity } D(\text{dB}) = -10\log_{10}\frac{P_4}{P_3} = -20\log_{10}|S_{43}|$$

$$\text{Isolation } L(\text{dB}) = -10\log_{10}\frac{P_4}{P_1} = -20\log_{10}|S_{41}|$$

$$\text{Return Loss } R(\text{dB}) = -10\log_{10}\frac{P_1{}'}{P_1} = -20\log_{10}|S_{11}|. \tag{4.76}$$

The couplers can be designed with different coupling factors. A number of microwave couplers are described below.

4.5.1   Two-Stub Branch Line Coupler

Figure 4.18 shows the design of a branch line coupler with two stubs (Pozar, 1990; Bhat and Koul, 1989). All transmission line sections have a length of $\lambda/4$. The S-parameters of the branch line coupler at the central frequency depend on the characteristic admittances of the line sections, Y_a and Y_b. The port admittance is taken to be Y_0.

For the input from Port 1,

$$S_{21} = -jY_0/Y_a, \quad S_{31} = -jY_b/Y_a \text{ and } S_{11} = S_{41} = 0. \tag{4.77}$$

The characteristic admittances of the line sections, Y_a and Y_b, for different coupling factors are tabulated in Table 4.1. The two-stub branch line coupler has a typical bandwidth of 10% measured with ±0.2 dB deviations.

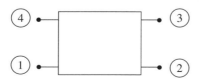

Figure 4.17   Four-port coupler

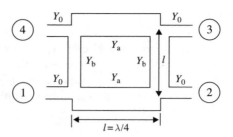

Figure 4.18   Two-stub branch line coupler

Table 4.1   Two-stub branch line coupler design

	S_{21}	S_{31}	Y_a/Y_0	Y_b/Y_0
3 dB 90° coupler	$-\dfrac{j}{\sqrt{2}}$	$-\dfrac{1}{\sqrt{2}}$	$\sqrt{2}$	1
6 dB 90° coupler	$-j\dfrac{\sqrt{3}}{2}$	$-\dfrac{1}{2}$	$\dfrac{2}{\sqrt{3}}$	$\dfrac{1}{\sqrt{3}}$
10 dB 90° coupler	$-j\dfrac{3}{\sqrt{10}}$	$-\dfrac{1}{\sqrt{10}}$	$\dfrac{\sqrt{10}}{3}$	$\dfrac{1}{3}$

4.5.2   Coupler with Flat Coupling Response

The coupler shown in Figure 4.19 that has a flat coupling response was first proposed by Riblet (1978). It is based on the design of a two-stub branch line coupler. Matching networks are introduced to all four ports to realise the flat coupling response. All transmission line sections and stubs have a length of $\lambda/4$ at the central frequency f_0. The characteristic admittances of the line sections depend on the requirements of power split ratio and bandwidth Δf. The design equations of the coupler are as follows:

$$K = \frac{\bar{Y}_a}{\bar{Y}_b} = \sqrt{\frac{|S_{21}|^2}{|S_{31}|^2} + 1} \tag{4.78}$$

$$\theta_0 = \cos^{-1}\left[\frac{1}{\sqrt{2}}\cos\left(\frac{\pi}{2}\left(1 + \frac{\Delta f}{2f_0}\right)\right)\right],$$

$$\frac{\sin^2\theta_0}{\bar{Y}_1} + \left(1 + \frac{\bar{Y}_2}{\bar{Y}_1}\right)^2 \cos^2\theta_0 = \frac{\sin\theta_0}{\bar{Y}_b\sqrt{K^2 - 1}},$$

$$\frac{\sin\theta_0}{\bar{Y}_1} - \left(1 + \frac{\bar{Y}_2}{\bar{Y}_1}\right)\left(\bar{Y}_1\sin\theta_0 - \bar{Y}_2\frac{\cos^2\theta_0}{\sin\theta_0}\right) = -\left(\frac{K+1}{K-1}\right)^2, \tag{4.79}$$

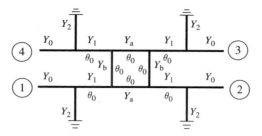

Figure 4.19 Coupler with flat coupling response

where

$$\bar{Y}_a = Y_a/Y_0, \quad \bar{Y}_b = Y_b/Y_0, \quad \bar{Y}_1 = Y_1/Y_0, \quad \bar{Y}_2 = Y_2/Y_0 \qquad (4.80)$$

and Y_0 is the port admittance. In these design equations, $\bar{Y}_b$ is an independent parameter which can be chosen so that the branch line impedances can be easily realised, and in most cases $\bar{Y}_b$ can be chosen to be close to 1.

4.5.3 Three-Stub Branch Line Coupler

Three-stub branch line couplers usually have larger bandwidth than two-stub branch line couplers with a typical bandwidth of 25%. The structure of a three-stub branch line coupler is shown in Figure 4.20 (Bhat and Koul, 1989). All line sections have a length of $\lambda/4$ at the central frequency. The S-parameters of the coupler depend on the characteristic admittance of the branch lines, i.e. Y_a, Y_b and Y_d. The port admittance is Y_0. For an input from Port 1, the S-parameters at the central frequency are

$$S_{11} = S_{41} = 0,$$

$$S_{21} = \frac{Y_a^2\left(Y_bY_d - Y_a^2\right)}{(Y_bY_d - Y_a^2)^2 + Y_d^2Y_0^2},$$

$$S_{31} = \frac{jY_0Y_a^2Y_d}{(Y_bY_d - Y_a^2)^2 + Y_d^2Y_0^2}, \qquad (4.81)$$

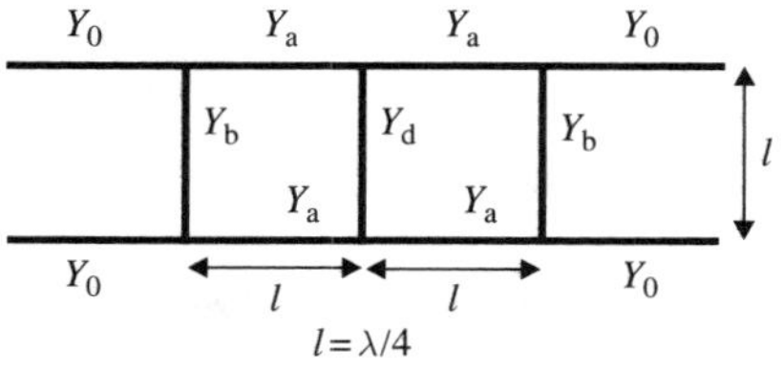

Figure 4.20 Three-stub branch line coupler

Table 4.2   Three-stub branch line coupler design

	S_{21}	S_{31}	Y_a	Y_b	Y_d
0 dB 90° coupler	0	$-j$	Y_0	Y_0	Y_0
3 dB 90° coupler	$\frac{1}{\sqrt{2}}$	$\frac{-j}{\sqrt{2}}$	$\sqrt{2}Y_0$	$\left(\sqrt{2}-1\right)Y_0$	$\sqrt{2}Y_0$

with

$$\left(Y_b^2 + Y_0^2\right)Y_d = 2Y_b Y_a^2. \tag{4.82}$$

Using the three-stub configuration, it is possible to design a 0 dB coupler. The required values of Y_a, Y_b and Y_d for the design are listed in Table 4.2. Similarly the parameters for a 3 dB coupler can also be obtained. They are listed in Table 4.2.

4.5.4   Coupled Line Couplers

Couplers can also be realised using coupled lines (Pozar, 1990; Bhat and Koul, 1989). The diagram of a single section backwave coupled line coupler is shown in Figure 4.21. The length of the coupled line section is $l = \lambda/4$ at the central frequency. With the definition of the voltage coupling factor, $C_v = (Z_{0e} - Z_{0o})/(Z_{0e} + Z_{0o})$, the voltages at Ports 3 and 2 normalised to the voltage at the input port, Port 1, are respectively

$$\left|\frac{V_3}{V_1}\right|^2 = \frac{C_v^2 \sin^2\theta}{1 - C_v^2\cos^2\theta}; \quad \left|\frac{V_2}{V_1}\right|^2 = \frac{1 - C_v^2}{1 - C_v^2\cos^2\theta}, \tag{4.83}$$

where $\theta = \beta l$ and $\theta = \pi/2$ at the central frequency. Port 4 is an isolated port.

At the central frequency, $|V_3/V_1|^2 = C_v^2$ and $|V_2/V_1|^2 = 1 - C_v^2$. For design purpose, the even and odd mode characteristic impedances of the coupled line are determined by the requirement of coupling strength C_v or $C_{v,dB} = 20\log_{10} C_v$ from

$$Z_{0e} = Z_0\sqrt{\frac{1+C_v}{1-C_v}}; \quad Z_{0o} = Z_0\sqrt{\frac{1-C_v}{1+C_v}}. \tag{4.84}$$

Figure 4.21   Coupled line coupler

The physical length of the line is determined by one of the two ways, i.e.

$$\frac{(\beta_e + \beta_o)}{2} l = \frac{\pi}{2} \quad \text{or} \quad \frac{2\beta_e \beta_o}{\beta_e + \beta_o} l = \frac{\pi}{2}. \tag{4.85}$$

The bandwidth of the coupled line coupler measured with $\pm 0.2\,\text{dB}$ derivation is typically 40%.

4.6 HYBRID RINGS

Hybrid rings are also the devices for the control of power flow. They are designed with transmission lines arranged in a ring format. Some designs are described below (Bhat and Koul, 1989).

4.6.1 Hybrid Ring Coupler

A hybrid ring that can be used as a coupler is shown in Figure 4.22 (Pon, 1961). The three line sections on the top half of the ring have a length of $\lambda/4$ at the central frequency, and the line section on the bottom half has a length of $3\lambda/4$. The port admittance of the four-port device is Y_0. The S-parameters of the hybrid ring coupler at the central frequency are

$$[S] = \begin{bmatrix} 0 & -j\bar{Y}_b & 0 & j\bar{Y}_a \\ -j\bar{Y}_b & 0 & -j\bar{Y}_a & 0 \\ 0 & -j\bar{Y}_a & 0 & -j\bar{Y}_b \\ j\bar{Y}_a & 0 & -j\bar{Y}_b & 0 \end{bmatrix} \tag{4.86}$$

with

$$\bar{Y}_a^2 + \bar{Y}_b^2 = 1 \tag{4.87}$$

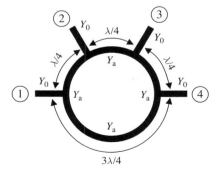

Figure 4.22 Hybrid ring coupler

where

$$\bar{Y}_a = Y_a/Y_0 \quad \text{and} \quad \bar{Y}_b = Y_b/Y_0 \tag{4.88}$$

are the normalised characteristic admittances of the line sections shown in Figure 4.22. When an input is made at Port 1, outputs at Ports 2 and 4 are out of phase by 180°, and Port 3 is isolated. When an input is made at Port 3, outputs at Ports 2 and 4 are in phase and Port 1 is isolated. The power splits depend on the values of Y_a and Y_b.

4.6.2   Rat-Race Hybrid

The rat-race hybrid is a special case of the hybrid ring described in Section 4.6.1, when $Y_a = Y_b = Y_0/\sqrt{2}$.

The S-parameters of the rat-race hybrid at the central frequency are

$$[S] = \begin{bmatrix} 0 & \frac{-j}{\sqrt{2}} & 0 & \frac{j}{\sqrt{2}} \\ \frac{-j}{\sqrt{2}} & 0 & \frac{-j}{\sqrt{2}} & 0 \\ 0 & \frac{-j}{\sqrt{2}} & 0 & \frac{-j}{\sqrt{2}} \\ \frac{j}{\sqrt{2}} & 0 & \frac{-j}{\sqrt{2}} & 0 \end{bmatrix}. \tag{4.89}$$

The rat-race hybrid is therefore a 3 dB coupler with good isolation between Ports 1 and 3 and between Ports 2 and 4 and useful bandwidth of typically 20%.

4.6.3   Wideband Rat-Race Hybrid

The bandwidth of the rat-race hybrid in Section 4.6.2 is limited by the properties of the $3\lambda/4$ section of the ring. A 3 dB rat-race hybrid with wideband response was proposed by March (1968) and is shown in Figure 4.23. The hybrid consists of three transmission line sections of length $\lambda/4$ and a coupled line section of the same length. Two opposite ends of the coupled line sections are short-circuited so that it effectively becomes a two-port device connected in the ring format. The characteristic impedances of the line sections for the 3dB hybrid are determined by

$$Z_r = 1.46Z_0; \quad Z_I = Z_r = \sqrt{Z_{0o}Z_{0e}}; \quad Z_{0e} = \left(1+\sqrt{2}\right)Z_r; \quad Z_{0o} = \left(\sqrt{2}-1\right)Z_r, \tag{4.90}$$

where Z_0 is the port impedance. The ±0.3 derivation bandwidth of the wideband hybrid coupler can be over an octave. The isolation is greater than 20 dB.

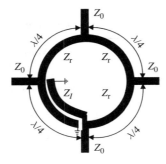

Figure 4.23 Wideband rat-race hybrid

4.6.4 Modified Hybrid Ring

The conversional hybrid ring coupler in Section 4.6.1 may be difficult to realise when the power split is large. A modified design to overcome this problem proposed by Agrawal and Mikucki (1986) is shown in Figure 4.24. The ring has six $\lambda/4$ transmission line sections with two values of characteristic impedance, Z_1 and Z_2. For different power split ratios, the values of Z_1 and Z_2 are tabulated in Table 4.3 for a port impedance of $Z_0 = 50\,\Omega$.

4.6.5 Modified Hybrid Ring with Improved Bandwidth

A modified hybrid ring with improved bandwidth is shown in Figure 4.25. The structure was proposed by Kim and Naito (1982). All line sections have a length of $\lambda/4$ at the central frequency. The optimum values of the characteristic admittance of the line sections for the improved bandwidth of a 3 dB hybrid ring coupler are

$$Y_1 = 0.88034Y_0, \qquad Y_2 = 1.34850Y_0, \qquad Y_3 = 3.57850Y_0,$$

$$Y_4 = 7.7350Y_0, \qquad Y_{1c} = 1.05570Y_0, \qquad Y_{2c} = 1.72860Y_0, , \qquad (4.91)$$

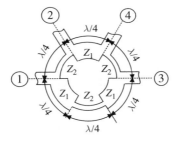

Figure 4.24 Modified hybrid ring

Table 4.3   The modified characteristics impedances of the lines for two hybrid-ring couplers (Agrawal and Mikucki, 1986)

Power-split ratio (dB)	Modified hybrid ring	
	$Z_1(\Omega)$	$Z_2(\Omega)$
0.0	70.7	70.7
1.0	72.3	66.9
2.0	74.5	63.8
3.0	77.3	61.1
4.0	80.8	59.0
5.0	84.9	57.1
6.0	89.8	55.6
7.0	95.6	54.4
8.0	102.3	53.5
9.0	109.9	52.7
10.0	118.6	52.1
11.0	128.4	51.6
12.0	139.4	51.2
13.0	151.9	50.9
14.0	165.9	50.7
15.0	181.6	50.5

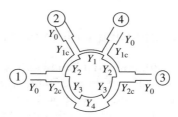

Figure 4.25   Modified hybrid ring with improved bandwidth

where Y_0 is the port admittance. The bandwidth of the optimised 3 dB coupler is 50.67%.

4.7   PHASE SHIFTERS

Phase shifters can be realised using ferrite, discrete or lumped inductors or capacitors, transmission lines and active components (Pozar, 1990). Two types of passive phase shifters are described below.

4.7.1   Transmission Line Phase Shifter

A transmission line phase shifter is made of a section of transmission line with characteristic impedance Z_0 which is the same as the port impedance, as shown in Figure 4.26. The S-parameters of the transmission line section of length L and phase constant β are given by

$$[S] = \begin{bmatrix} 0 & e^{-j\theta} \\ e^{-j\theta} & 0 \end{bmatrix} \tag{4.92}$$

where $\theta = \beta L$. Hence the transmission between two ports has a phase delay of θ, and both ports are fully matched.

4.7.2   LC Phase Shifters

LC phase shifters can be made of discrete inductor and capacitor components connected in 'T', 'π', to 'L' networks. For an 'L' network phase shifter as shown in Figure 4.27 with normalised impedance and admittance quantities x and b, the S-parameters are

$$[S] = \frac{\begin{bmatrix} \dfrac{-xb + j(x - b)}{2} & 2 \\ 2 & -xb + j(x - b) \end{bmatrix}}{(2 - xb) + j(x + b)}. \tag{4.93}$$

The phase shifter has an insertion loss of

$$|S_{21}| = \frac{2}{\sqrt{(2 - xb)^2 + (x + b)^2}} \tag{4.94}$$

and a phase shift

$$\theta = \arctan\left(\frac{x + b}{2 - xb}\right) \tag{4.95}$$

It is noted that such a phase shifter is not fully matched at both ports. The reflections at the ports are

$$|S_{11}| = |S_{22}| = \sqrt{\frac{(xb)^2 + (x - b)^2}{(2 - xb)^2 + (x + b)^2}}. \tag{4.96}$$

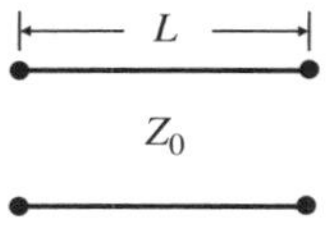

Figure 4.26   Transmission line phase shifter

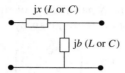

Figure 4.27 'L' network discrete element phase shifter

In the special case that $x = 0$, the insertion loss and phase shift are, respectively,

$$|S_{21}| = \frac{1}{\sqrt{1 + b^2/4}} \qquad (4.97)$$

and

$$\theta = \arctan(b/2). \qquad (4.98)$$

4.8   FILTERS

4.8.1   Maximally Flat Response

The insertion loss of a low-pass filter with maximally flat response has the following form (Pozar, 1990; Matthaei, Young and Jones, 1965):

$$P_{LR} = 1 + \omega_1^{2N}, \qquad (4.99)$$

where N is the order of the filter and the 3 dB cut-off frequency is at $\omega_c = 1$. The insertion loss is shown in Figure 4.28 for different values of N.

4.8.2   Chebyshev Response

The insertion loss of a low-pass filter with Chebyshev response or equal ripples in the pass band has the following form (Pozar, 1990; Matthaei, Young and Jones, 1965):

$$P_{LR} = 1 + k^2 T_N^2 (\omega), \qquad (4.100)$$

where $1 + k^2$ is the pass band ripple level, $T_N (\omega)$ is a Chebyshev polynomial of order N and

$$T_N (x) = \cos \left(N \cos^{-1} x \right) \text{ for } |x| < 1 \quad \text{and} \quad T_N (x) = \cosh \left(N \cosh^{-1} x \right) \text{ for } |x| > 1$$
$$(4.101)$$

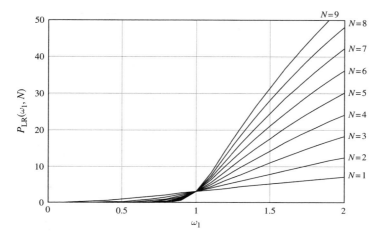

Figure 4.28   Maximally flat response $P_{LR}(\omega_1, N)$

with

$$T_N(0) = \begin{cases} 0 \text{ for } N \text{ odd} \\ 1 \text{ for } N \text{ even} \end{cases} \text{ and } T_N(1) = 1. \tag{4.102}$$

The cut-off frequency of the low-pass filter is at $\omega_c = 1$. The insertion loss is shown in Figure 4.29 for different values of N.

4.8.3   Maximally Flat Low-Pass Filters with $\omega_c = 1$

The circuits of maximally flat low-pass filters with $\omega_c = 1$ for a generator impedance $R_0 = 1\,\Omega$ and load resistance $R_{N+1} = 1\,\Omega$ are shown in Figure 4.30(a) and (b), respectively, for a prototype beginning with a shunt capacitor and a prototype beginning with a series inductor. A total of N capacitors or inductors are used for a filter of order N. The element values, g_k, for both prototypes are given by (Matthaei, Young and Jones, 1965)

$$g_0 = 1,$$

$$g_k = 2 \sin\left(\frac{(2k-1)\pi}{2N}\right) \text{ for } k = 1, 2, \cdots, N,$$

$$g_{N+1} = 1 \tag{4.103}$$

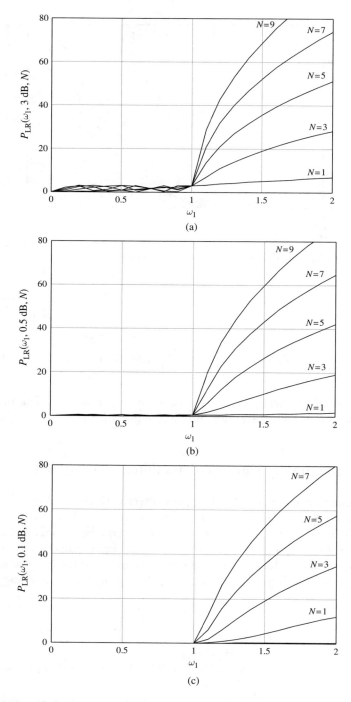

Figure 4.29   Chebyshev response $P_{LR}(\omega_1,\text{Ripple Level},N)$: (a) 3 dB ripple level, (b) 0.5 dB ripple level and (c) 0.1 dB ripple level

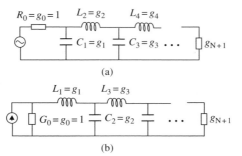

Figure 4.30 Filter prototypes: (a) prototype beginning with a capacitor and (b) prototype beginning with an inductor

Table 4.4 g_k values for maximally flat low-pass filters with $g_0 = 1$, $g_{N+1} = 1$ and $\omega_c = 1$ (Matthaei, Young and Jones, 1965)

N	g_1	g_2	g_3	g_4	g_5
1	2.0000				
2	1.4142	1.4142			
3	1.0000	2.0000	1.0000		
4	0.7654	1.8478	1.8478	0.7654	
5	0.6180	1.6180	2.0000	1.6180	0.6180

and g_k for $k = 1, 2, \ldots, N$ is an inductance value for a series inductor and a capacitance value for a shunt capacitor. The values of g_k for $N = 1 - 5$ are tabulated in Table 4.4.

4.8.4 Chebyshev Low-Pass Filters with $\omega_c = 1$

The circuits of Chebyshev low-pass filters with $\omega_c = 1$ for a generator impedance $R_0 = 1\,\Omega$ and load resistance $R_{N+1} = 1\,\Omega$ are shown similar to those shown in Figure 4.30(a) and (b). However, for Chebyshev low-pass filters with $R_{N+1} = 1\,\Omega$, the order N must be an odd number, i.e., $N = 1, 3, 5, \ldots$. Once N is determined, both prototypes consist of N inductors or capacitors. The element values, g_k, for both prototypes are given by (Matthaei, Young and Jones, 1965)

$$g_0 = 1,$$

$$g_1 = 2a_1/\gamma_{AR},$$

$$g_k = \frac{4a_{k-1}a_k}{b_{k-1}g_{k-1}} \text{ for } k = 2, 3, \ldots, N,$$

$$g_{N+1} = 1 \text{ (for odd } N \text{ only),} \tag{4.104}$$

where

$$a_k = \sin\left[\frac{(2k-1)\,\pi}{2N}\right] \quad \text{for } k = 1, 2, \ldots, N,$$

$$b_k = \gamma_{AR}^2 + \sin^2\left(\frac{k\pi}{N}\right) \quad \text{for } k = 1, 2, \ldots, N,$$

$$\gamma_{AR} = \sinh\left(\frac{\beta_{AR}}{2N}\right),$$

$$\beta_{AR} = \ln\left(\coth\frac{L_{AR}}{17.37}\right),$$

$$L_{AR} = 10\log_{10}(1 + k^2) = \text{ripple level in dB.} \tag{4.105}$$

The element value g_k can be an inductance value for a series inductor or a capacitive value for a shunt capacitor. The values of g_k for $N = 1, 3, 5, 7, 9$ and ripple level of 0.5, 0.1 and 3 dB are tabulated in Table 4.5.

4.8.5 Filter Transformations

The low-pass filter prototypes described in Sections 4.8.3 and 4.8.4 are for the case that $R_0 = 1\,\Omega$, $R_{N+1} = 1\,\Omega$ and the cut-off frequency is $\omega_c = 1$. These low-pass prototypes can be readily transformed to a low-pass filter (LPF) with a cut-off frequency ω_{LP}, a high-pass filter (HPF) with a high-pass

Table 4.5 g_k values for Chebyshev low-pass filters with $g_0 = 1$, $g_{N+1} = 1$ and $\omega_c = 1$ (Matthaei, Young and Jones, 1965)

N	g_1	g_2	g_3	g_4	g_5	g_6	g_7	g_8	g_9
Ripple = 0.5 dB									
1	0.6986								
3	1.5963	1.0967	1.5963						
5	1.7058	1.2296	2.5408	1.2296	1.7058				
7	1.7372	1.2583	2.6381	1.3444	2.6381	1.2583	1.7372		
9	1.7504	1.2690	2.6678	1.3673	2.7239	1.3673	2.6678	1.2690	1.7504
Ripple = 0.1 dB									
1	0.3052								
3	1.0315	1.1474	1.0315						
5	1.1468	1.3712	1.9750	1.3712	1.1468				
7	1.1811	1.4228	2.0966	1.5733	2.0966	1.4228	1.1811		
9	1.1956	1.4425	2.1345	1.6167	2.2053	1.6167	2.1345	1.4425	1.1956
Ripple = 3 dB									
1	1.9953								
3	3.3487	0.7117	3.3487						
5	3.4817	0.7618	4.5381	0.7618	3.4817				
7	3.5182	0.7723	4.6386	0.8039	4.6386	0.7723	3.5182		
9	3.5340	0.7760	4.6692	0.8118	4.7272	0.8118	4.6692	0.7760	3.5340

Table 4.6 Frequency projections

LPF	HPF	BPF	BSF
$\omega_1 = \dfrac{\omega}{\omega_{LP}}$	$\omega_1 = -\dfrac{\omega_{HP}}{\omega}$	$\omega_1 = \dfrac{1}{\Delta}\left(\dfrac{\omega}{\omega_0} - \dfrac{\omega_0}{\omega}\right)$	$\omega_1 = \Delta\left(\dfrac{\omega}{\omega_0} - \dfrac{\omega_0}{\omega}\right)^{-1}$
$\omega = 2\pi f$	$\omega = 2\pi f$	$\Delta = \dfrac{f_2 - f_1}{f_0}$	$\Delta = \dfrac{f_2 - f_1}{f_0}$
$\omega_{LP} = 2\pi f_{LP}$	$\omega_{HP} = 2\pi f_{HP}$	$\omega_0 = 2\pi f_0$	$\omega_0 = 2\pi f_0$
		$\omega = 2\pi f$	$\omega = 2\pi f$

band-edge frequency ω_{HP}, a bandpass filter (BPF) at a central frequency f_0 with bandwidth $(f_2 - f_1)$, a bandstop filter (BSF) at a central frequency f_0 with bandwidth $(f_2 - f_1)$ and in all cases above with $R_0 = R_{N+1} = Z_0$.

The frequency projections for these transformations are tabulated in Table 4.6 (Pozar, 1990; Bahl, 1997). These projections are used to determine the order of the filter needed to meet the attenuation requirement at frequencies outside the band of interest.

For the transformation to a low-pass filter, the series element g_k is replaced by an inductor L_k and the shunt element g_k by a capacitor C_k.

For the transformation to a high-pass filter, the series element g_k is replaced by a series capacitor C_k and the shunt element g_k by an inductor L_k.

For the transformation to a bandpass filter, the series element g_k is replaced by an inductor L_k and a capacitor C_k connected in series. The shunt element g_k is replaced by an inductor L_k and a capacitor C_k connected in parallel.

For the transformation to a bandstop filter, the series element g_k is replaced by an inductor L_k and a capacitor C_k connected in parallel. The shunt element g_k is replaced by an inductor L_k and a capacitor C_k connected in series.

The values of L_k and C_k for these transformations are given in Table 4.7 (Pozar, 1990; Bahl, 1997).

4.8.6 Step Impedance Low-Pass Filters

A step impedance low-pass filter is made of short transmission line sections of high or low characteristic impedance, e.g., $Z_{high} = 150\,\Omega$ and $Z_{low} = 10\,\Omega$, as shown in Figure 4.31 (Pozar, 1990). The low-pass filter is designed in the usual way using the procedures described in Sections 4.8.1–4.8.5. But the series inductor L_k is replaced by a high-impedance line section with a length l_k given by

Table 4.7 Filter transformation (Pozar, 1990)

Element	LPF	HPF	BPF	BSF
Series g_k	Series L $L_k = g_k \dfrac{Z_0}{\omega_{LP}}$	Series C $C_k = \dfrac{1}{g_k \omega_{HP} Z_0}$	Series-tuned series elements $L_k = g_k \dfrac{Z_0}{\omega_0 \Delta}$ $C_k = \dfrac{\Delta}{g_k Z_0 \omega_0}$	Shunt-tuned series elements $L_k = g_k \dfrac{\Delta Z_0}{\omega_0}$ $C_k = \dfrac{1}{g_k \Delta \omega_0 Z_0}$
Shunt g_k	Shunt C $C_k = g_k \dfrac{1}{\omega_{LP} Z_0}$	Shunt L $L_k = \dfrac{Z_0}{g_k \omega_{HP}}$	Shunt-tuned shunt elements $L_k = \dfrac{\Delta Z_0}{g_k \omega_0}$ $C_k = \dfrac{g_k}{\omega_0 \Delta Z_0}$	Series-tuned shunt elements $L_k = \dfrac{Z_0}{g_k \Delta \omega_0}$ $C_k = g_k \dfrac{\Delta}{\omega_0 Z_0}$

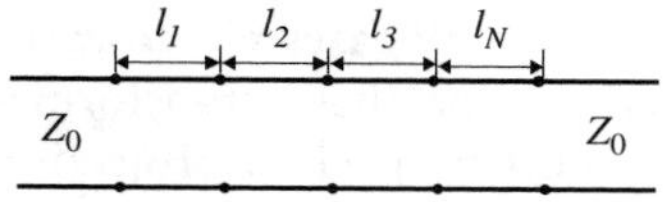

Figure 4.31 Step impedance low-pass filter

$$\beta l_k = \frac{L_k Z_0}{Z_{\text{high}}}, \tag{4.106}$$

where β is the phase constant at the central frequency. The shunt capacitor C_k is replaced by a low-impedance line with a length l_k given by

$$\beta l_k = \frac{C_k Z_{\text{low}}}{Z_0}. \tag{4.107}$$

In all cases, $\beta l < \pi/4$.

4.8.7 Bandpass and Bandstop Filters Using $\lambda/4$ Resonators

At microwave frequencies, bandpass and bandstop filters can be realised using $\lambda/4$ resonators as shown in Figure 4.32 (Pozar, 1990). The length of the line section or the spacing l between the stubs is also $\lambda/4$, so that $\theta = \beta l = \pi/2$. If the stubs are short-circuited at their ends, the filter has a bandpass response. If they are open-circuited, the filter has a bandstop response. The required characteristic impedances for the bandpass response are

$$Z_{0n} = \frac{\pi Z_0 \Delta}{4 g_n} \text{ for } n = 1 \text{ to N}, \tag{4.108}$$

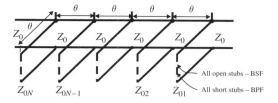

Figure 4.32 Bandpass and bandstop filters using $\lambda/4$ resonators

where Z_0 is the source impedance and also the characteristic impedance of the line sections connecting the stubs, and Δ is the relative bandwidth given in Table 4.6.

Similarly those for bandstop response are given by

$$Z_{0n} = \frac{4Z_0}{\pi g_n} \text{ for } n = 1 \text{ to } N. \tag{4.109}$$

For equal-ripple designs, N must be an odd number.

4.8.8 Bandpass Filters Using $\lambda/4$ Connecting Lines and Short-Circuited Stubs

Figure 4.33 shows an alternative design of bandpass filters using $\lambda/4$ connecting lines and short-circuited stubs (Pozar, 1990; Bhat and Koul, 1989; Matthaei, Young and Jones, 1965). All line sections are $\lambda/4$ long at the central frequency. The procedures for designing this type of filter are as follows:

(1) Map the frequency to ω_1 domain

$$\omega_1 = \frac{2}{\Delta}\left(\frac{\omega - \omega_0}{\omega_0}\right); \quad \Delta = \frac{f_2 - f_1}{f_0}, \quad f_0 = \frac{f_1 + f_2}{2}, \quad \omega = 2\pi f, \quad \omega_0 = 2\pi f_0, \tag{4.110}$$

where f_0 is the central frequency and $(f_2 - f_1)$ defines the bandwidth.

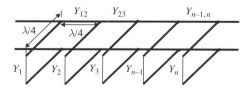

Figure 4.33 Bandpass filters using $\lambda/4$ connecting lines and short-circuited stubs

(2) Use the attenuation requirements and low-pass filter response curves in Figure 4.28 or 4.29 to determine the order N required, where N is also the number of stubs in the filter.

(3) Obtain $g_1, g_2, \ldots, g_N$ values for maximally flat or Chebyshev ($N =$ odd) response.

(4) Calculate length, i.e. $\lambda/4$, at f_0.

(5) Compute the characteristic admittances of connecting lines and stubs:

$$\theta_1 = \frac{\pi}{2}\left(1 - \frac{\Delta}{2}\right) = \frac{\pi}{2}\frac{f_1}{f_0}$$

$C_a = 2dg_1$ where $d = 0.5 - 1$ but typically $d = 1$ is chosen,

$$\frac{J_{12}}{Y_0} = \left(\frac{C_a}{g_2}\right)^{1/2}; \quad \frac{J_{k,k+1}}{Y_0} = \frac{g_0 C_a}{\sqrt{g_k g_{k+1}}}, \quad k = 2 \text{ to } N-2,$$

$$\frac{J_{N-1,N}}{Y_0} = \left(\frac{C_a}{g_{N-1}}\right)^{1/2},$$

$$N_{k,k+1} = \left[\left(\frac{J_{k,k+1}}{Y_A}\right)^2 + \left(\frac{C_a \tan\theta_1}{2}\right)^2\right]^{1/2}, \quad k = 1 \text{ to } N-1,$$

$$Y_1 = Y_0 (1-d) g_1 \tan\theta_1 + Y_0\left(N_{12} - \frac{J_{12}}{Y_0}\right),$$

$$Y_k|_{k=2 \text{ to } (N-1)} = Y_0\left(N_{k-1,k} + N_{k,k+1} - \frac{J_{k-1,k}}{Y_0} - \frac{J_{k,k+1}}{Y_0}\right),$$

$$Y_N = Y_0 (g_N - dg_1) \tan\theta_1 + Y_0\left(N_{N-1,N} - \frac{J_{N-1,N}}{Y_0}\right),$$

$$Y_{k,k+1} = J_{k,k+1} \text{ (for } k = 1 \text{ to } N-1),$$ (4.111)

where Y_0 is the source admittance and load conductance.

4.8.9   Coupled Line Bandpass Filters

Bandpass filters can also be realised using coupled lines (Pozar, 1990; Bhat and Koul, 1989; Matthaei, Young and Jones, 1965; Cohn, 1958). A diagram of the coupled line bandpass filter is shown in Figure 4.34. At the central frequency, the coupled line sections are $\lambda/4$ in length, where

$$\lambda = \frac{2\lambda_{\text{air}}}{\left(\sqrt{\varepsilon_{\text{re}}} + \sqrt{\varepsilon_{\text{ro}}}\right)}$$

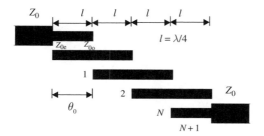

Figure 4.34 Coupled line bandpass filters in microstrip configuration

with $\lambda_{\text{air}} = c/f$ taking the propagation of both even and odd modes into account. The design procedures of the coupled line filter are as follows:

(1) Map the frequency to ω_1 domain:

$$\omega_1 = \frac{1}{\Delta}\left(\frac{\omega - \omega_0}{\omega_0}\right), \quad \Delta = \frac{f_2 - f_1}{f_0}, \quad f_0 = \frac{f_1 + f_2}{2}, \quad \omega = 2\pi f, \quad \omega_0 = 2\pi f_0., \quad (4.112)$$

(2) Use the attenuation requirements and low-pass filter response curves in Figure 4.28 or 4.29 to determine the order N required. The number of coupled line sections will be $(N+1)$.

(3) Obtain $g_1, g_2, \ldots, g_N$ values for maximally flat or Chebyshev ($N = $ odd) response.

(4) Calculate couple line length, i.e. $\lambda/4$, at f_0.

(5) Compute Z_{0e} and Z_{0o} of the coupled lines:

$$Z_{0ek} = Z_0\left[1 + J_k Z_0 + (J_k Z_0)^2\right],$$

$$Z_{0ok} = Z_0\left[1 - J_k Z_0 + (J_k Z_0)^2\right] \text{ for } k = 1, 2, \ldots, N+1, \quad (4.113)$$

where

$$J_1 Z_0 = \sqrt{\frac{\pi\Delta}{2g_1}},$$

$$J_2 Z_0 = \frac{\pi\Delta}{2\sqrt{g_1 g_2}},$$

$$\cdots\cdots$$

$$J_k Z_0 = \frac{\pi \Delta}{2\sqrt{g_{k-1}g_k}},$$

$$Z_0 J_{N+1} = \sqrt{\frac{\pi \Delta}{2g_N}} \qquad (4.114)$$

and Z_0 is the source and load impedance.

4.8.10   End-Coupled Resonator Filters

End-coupled resonator filters, as shown in Figure 4.35, also have bandpass response (Pozar, 1990; Bhat and Koul, 1989). The design procedures for the type of filter are as follows:

(1) Map the frequency to ω_1 domain:

$$\omega_1 = \frac{1}{\Delta}\left(\frac{\omega - \omega_0}{\omega_0}\right), \quad \Delta = \frac{f_2 - f_1}{f_0}, \quad f_0 = \frac{f_1 + f_2}{2}, \quad \omega = 2\pi f, \quad \omega_0 = 2\pi f_0.$$

$$(4.115)$$

(2) Use the attenuation requirements and low-pass filter response curves in Figure 4.28 or 4.29 to determine the order N required. The filter will have N resonator sections and $(N+1)$ gaps.
(3) Obtain $g_1, g_2, \ldots, g_N$ values for maximally flat or Chebyshev ($N =$ odd) response.
(4) Calculate gap capacitance C_i:

$$C_i = \frac{B_i}{\omega_0}, \qquad (4.116)$$

where

$$B_i = \frac{(Z_0 J_i)Z_0^{-1}}{1 - (Z_0 J_i)^2}, \quad i = 1, \ldots, (N+1),$$

$$Z_0 J_1 = \sqrt{\frac{\pi}{2g_1}},$$

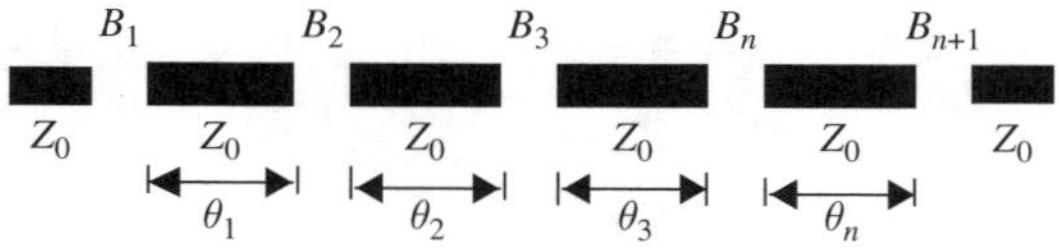

Figure 4.35   End-coupled resonator filters in microstrip configuration

$$Z_0 J_k = \frac{\pi}{2\sqrt{g_{k-1}g_k}} \text{ for } k = 2, 3, \ldots, N,$$

$$Z_0 J_{N+1} = \sqrt{\frac{\pi}{2g_N}}. \tag{4.117}$$

(5) Calculate resonator length l_i:

$$\theta_i = \beta l_i = \pi - \frac{1}{2}\left[\tan^{-1}(2Z_0 B_i) + \tan^{-1}(2Z_0 B_{i+1})\right]. \tag{4.118}$$

Again Z_0 is the source and load impedance.

4.9 AMPLIFIER DESIGN

4.9.1 Maximum Gain Amplifier Design

For an amplifier with matching circuits at the input and output ends of the transistor as shown in Figure 4.36, the transducer gain of the network of three blocks is (Pozar, 1990)

$$G_T = G_s G_0 G_L = \left(\frac{1 - |\Gamma_s|^2}{|1 - \Gamma_{in}\Gamma_s|^2}\right)(|S_{21}|^2)\left(\frac{1 - |\Gamma_L|^2}{|1 - S_{22}\Gamma_L|^2}\right), \tag{4.119}$$

where Γ_s, Γ_L, Γ_{in} and Γ_{out} are the reflection coefficients shown in Figure 4.36. The reflection coefficients Γ_{in} and Γ_{out} are related to the S-parameters of the transistor Γ_s and Γ_L by

$$\Gamma_{in} = S_{11} + \frac{S_{12}S_{21}\Gamma_L}{1 - S_{22}\Gamma_L} \quad \text{and} \quad \Gamma_{out} = S_{22} + \frac{S_{12}S_{21}\Gamma_s}{1 - S_{22}\Gamma_s}. \tag{4.120}$$

The unconditionally stable conditions of the amplifier are

$$K = \frac{1 - |S_{11}|^2 - |S_{22}|^2 + |\Delta|^2}{2|S_{12}S_{21}|} > 1 \quad \text{and} \quad |\Delta| = |S_{11}S_{22} - S_{12}S_{21}| < 1. \tag{4.121}$$

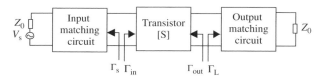

Figure 4.36 Amplifier design

The design of an amplifier with maximum gain (Pozar, 1990) requires that

$$\Gamma_{in} = \Gamma_s^* \quad \text{and} \quad \Gamma_{out} = \Gamma_L^* \tag{4.122}$$

so that

$$\Gamma_s = \frac{B_1 \pm \left(B_1^2 - 4|C_1|^2\right)^{1/2}}{2C_1}, \quad \Gamma_L = \frac{B_2 \pm \left(B_2^2 - 4|C_2|^2\right)^{1/2}}{2C_2}, \tag{4.123}$$

where

$$B_1 = 1 + |S_{11}|^2 - |S_{22}|^2 - ||^2$$
$$B_2 = 1 + |S_{22}|^2 - |S_{11}|^2 - ||^2$$
$$C_1 = S_{11} - S_{22}^*$$
$$\Delta = S_{11}S_{22} - S_{12}S_{21}. \tag{4.124}$$

The maximum transducer gain is then

$$G_T = G_{Tmax} = \frac{1}{\left(1 - |\Gamma_s|^2\right)} |S_{21}|^2 \frac{\left(1 - |\Gamma_L|^2\right)}{|1 - S_{22}\Gamma_L|^2}. \tag{4.125}$$

For the purpose of impedance matching in the amplifier, the admittance transformation from normalised load admittance $y_L = 1 + jb_L$ to normalised input admittance $y_{in} = g_{in} + jb_{in}$ using a transmission line section of characteristic impedance Z_0 and a stub of the same characteristic impedance, i.e. Z_0, as shown in Figure 4.37 requires that

$$b_L = \sqrt{\frac{(1 - g_{in})^2 + b_{in}^2}{g_{in}}} = \tan(\beta L_{stub}) \tag{4.126}$$

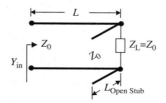

Figure 4.37 Amplifier impedance matching using an open-circuit stub and a transmission line section

and

$$\tan(\beta L) = \frac{(b_{in} - b_L)(1 - g_{in} + b_{in}b_L) - (1 - g_{in})(b_{in} + g_{in}b_L)}{(1 - g_{in} + b_{in}b_L)^2 + (b_{in} + g_{in}b_L)^2}, \quad (4.127)$$

where L and L_{stub} are the lengths of the transmission line and open-circuited stub, respectively, and β is the phase constant. These two equations can be used to determine the length of the open-circuited stub connected in parallel to the load or source impedance of the amplifier circuit and the length of the transmission line section required for impedance matching at both input and output ends of the transistor.

The impedance matching of the amplifier can also be realised using discrete L and C components (Radmanesh, 2001; Rudwig and Bretchko, 2000). For the requirement of the normalised impedance $z_{in} = r_{in} + jx_{in}$ with $r_{in} < 1$, the impedance transformation circuit as shown in Figure 4.38(a) can be used. The required circuit elements are

$$X_p/Z_0 = \pm\sqrt{\frac{r_{in}}{1 - r_{in}}} \quad \text{and} \quad X_s/Z_0 = (x_{in} \mp \sqrt{r_{in}(1 - r_{in})}). \quad (4.128)$$

Otherwise for the requirement of the normalised admittance $y_{in} = g_{in} + jb_{in}$ with $g_{in} < 1$, the admittance transformation circuit as shown in Figure 4.38(b) can be used. The required circuit elements are

$$X_s/Z_0 = \pm\sqrt{\frac{1 - g_{in}}{g_{in}}} \quad \text{and} \quad X_p/Z_0 = -(b_{in} \pm \sqrt{g_{in}(1 - g_{in}))^{-1}}. \quad (4.129)$$

4.9.2 Broadband Amplifier Design

The amplifier discussed in Section 4.9.1 can achieve maximum gain, but it is classified as a narrow band amplifier. The bandwidth of the amplifier can be much improved using the balanced amplifier configuration, as shown in Figure 4.39(a) (Radmanesh, 2001). The balanced amplifier consists of two maximum gain amplifiers designed as in Section 4.9.1 and two 3 dB couplers. The first coupler splits the input signal into two parts with a

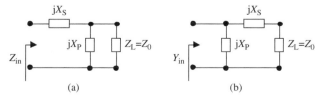

Figure 4.38 Amplifier impedance matching circuits for (a) $r_{in} < 1$ and (b) $g_{in} < 1$

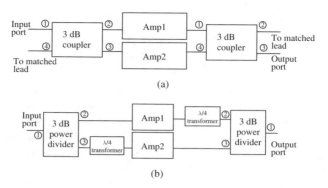

Figure 4.39 Balanced amplifier using (a) 3 dB couplers and (b) 3 dB power dividers

phase difference of 90°. The signals are amplified by Amp1 and Amp2, respectively, and combined at the output. In the case that the two amplifiers, Amp1 and Amp2, are identical, the signals combined at the output are in phase at the central design frequency. The S-parameters of the balanced amplifier in terms of power transfer at the central design frequency f_0 are

$$|S(f_0)|^2 = \begin{bmatrix} 0 & |S_{21,\mathrm{amp}}(f_0)|^2 \\ |S_{12,\mathrm{amp}}(f_0)|^2 & 0 \end{bmatrix}, \tag{4.130}$$

where the subscript amp refers to the value for a single amplifier Amp1 or Amp2. The power transfer of the balanced amplifier is the same as a single amplifier Amp1 or Amp2. But it has very good impedance matching characteristics at both input and output ports.

The 3 dB couplers can be made of branch line couplers. They may also be replaced by Wilkinson power dividers together with two quarter-wave transformers (Rudwig and Bretchko, 2000). The first power divider splits the input power into two equal amounts of the same phase. One of the quarter-wave transformers is used to shift the phase of a signal in one branch by 90° at the central frequency before amplification, and the other transformer in the other branch by the same phase angle after amplification. Both signals are then combined in phase by the second power divider.

4.9.3 High-Frequency Small Signal FET Circuit Model

At microwave frequencies, the characteristics of active devices are usually described by their S-parameters. However, equivalent circuits have also been used to represent active devices, particularly for small signal operations (Weber, 2001; Soares, 1988). An equivalent circuit for a FET operating under small signal condition is shown in Figure 4.40 as an example. The

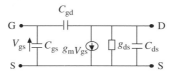

Figure 4.40   Equivalent circuit for FET

S-parameters of the equivalent circuit for a common source configuration with respective port admittance Y_0 are given by

$$S_{11} = \frac{(Y_0 - Y_{gg})(Y_0 + Y_{dd}) + Y_{gd}Y_{dg}}{(Y_0 + Y_{gg})(Y_0 + Y_{dd}) - Y_{gd}Y_{dg}}; \quad S_{12} = \frac{-2Y_{gd}Y_0}{(Y_0 + Y_{gg})(Y_0 + Y_{dd}) - Y_{gd}Y_{dg}};$$

$$S_{21} = \frac{-2Y_{dg}Y_0}{(Y_0 + Y_{gg})(Y_0 + Y_{dd}) - Y_{gd}Y_{dg}}; \quad S_{22} = \frac{(Y_0 + Y_{gg})(Y_0 - Y_{dd}) + Y_{gd}Y_{dg}}{(Y_0 + Y_{gg})(Y_0 + Y_{dd}) - Y_{gd}Y_{dg}},$$

$$(4.131)$$

where (Soares, 1988)

$$Y_{gg} = j\omega(C_{gs} + C_{gd}); \quad Y_{gd} = -j\omega C_{gd}; \quad Y_{dg} = g_m - j\omega C_{gd}; \quad Y_{dd} = g_{ds} + j\omega(C_{ds} + C_{gd}).$$

$$(4.132)$$

Some typical circuit element values are $C_{gs} = 0.5$ pF, $C_{gd} = 0.05$ pF, $C_{ds} = 0.1$ pF, $g_{gd} = 5$ mS and $g_m = 50$ mS.

4.9.4   Negative Feedback Amplifier Design

An amplifier with negative feedback can have flat gain response over a very wide frequency band. Such an amplifier also has good input and output impedance matching characteristics. It is used when broadband amplifiers such as the balanced amplifier described in Section 4.9.2 fails to meet bandwidth requirement. An amplifier with negative feedback using FET is shown in Figure 4.41(a). It consists of a series element R_1 and a shunt element R_2. The equivalent circuit of the amplifier is shown in Figure 4.41(b). The S-parameters of the circuit with respective port impedance Z_0 are given by (Radmanesh, 2001)

$$[S] = \begin{bmatrix} 1 - \dfrac{g_m Z_0^2}{R_2(1 + g_m R_1)} & \dfrac{2Z_0}{R_2} \\ \dfrac{2Z_0}{R_2} - \dfrac{2g_m Z_0}{(1 + g_m R_1)} & 1 - \dfrac{g_m Z_0^2}{R_2(1 + g_m R_1)} \end{bmatrix} \left[1 + \dfrac{2Z_0}{R_2} + \dfrac{g_m Z_0^2}{R_2(1 + g_m R_1)} \right]^{-1}.$$

$$(4.133)$$

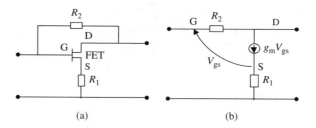

Figure 4.41 (a) Negative feedback amplifier and (b) its equivalent circuit

If both input and output ports are required to be matched so that $S_{11} = S_{22} = 0$, the minimum requirement for g_m is that

$$g_m \geq \frac{R_2}{Z_0^2}. \tag{4.134}$$

In this case,

$$S_{21} = \frac{Z_0 - R_2}{Z_0} \quad \text{and} \quad S_{12} = \frac{Z_0}{R_2 + Z_0}, \tag{4.135}$$

which depend on R_2 and Z_0 only.

If the characteristics of FET is described by S-parameters, the connection of FET with a series resistor R_1 as shown in Figure 4.42(a) can be taken as a series-to-series connection of two networks as shown in Figure 4.42(b). The resultant S-parameters can be obtained as

$$[S^{(FET,R_1)}] = -h([-[S^{(FET)}], -[S^{(R_1)}]), \tag{4.136}$$

where

$$[S^{(R_1)}] = \begin{bmatrix} -1 & \dfrac{2R_1}{Z_0} \\ \dfrac{2R_1}{Z_0} & -1 \end{bmatrix} \left[1 + \frac{2R_1}{Z_0} \right]^{-1} \tag{4.137}$$

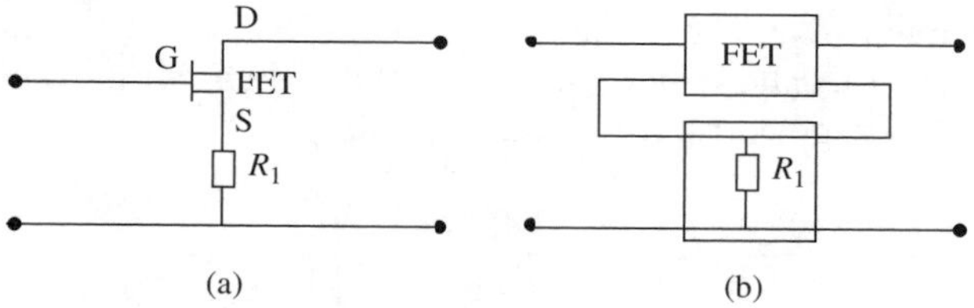

Figure 4.42 (a) FET with a resistor connected to its source terminal and (b) its network presentation

and the function $h([S_1],[S_2])$ has been given in section 1.6.3. The FET-R_1 combined network is again in a parallel-to-parallel connection with the R_2 network. The S-parameters of the R_2 network are

$$[S^{(R_2)}] = \frac{\begin{bmatrix} R_2 & 2Z_0 \\ 2Z_0 & R_2 \end{bmatrix}}{R_2 + 2Z_0}. \tag{4.138}$$

The S-parameters of the amplifier are therefore

$$[S] = h([[S^{(\text{FET},R_1)}],[S^{(R_1)}]]). \tag{4.139}$$

REFERENCES

Agrawal, A.K. and Mikucki, G.F. (1986) 'A printed circuit hybrid ring directional coupler for arbitrary power division', *IEEE Transactions on Microwave Theory and Techniques*, **MTT-34**, 1401–7.

Bahl, I.J. (1997) 'Chapter 3: Filters, hybrids, and couplers, power combiners and matching networks', in *Handbook of Microwave and Optical Components – Vol. 1: Microwave Passive and Antenna Components* (ed. K. Chang), John Wiley & Sons, Ltd, New York, pp 118–90.

Bhat, B. and Koul, S.K. (1989) *Stripline-Like Transmission Lines for Microwave Integrated Circuits*, John Wiley & Sons, Ltd, New York.

Cohn, S.B. (1958) 'Parallel-coupled transmission line resonator filter', *IEEE Transactions on Microwave Theory and Techniques*, **MTT-6**, 223–31.

Cohn, S.B. (1968) 'A class of broadband three-port TEM-mode hybrids', *IEEE Transactions on Microwave Theory and Techniques*, **MTT-16**, 110–6.

Collin, R.E. (1966) *Foundations for Microwave Engineering*, McGraw-Hill, New York.

Ginzton, E.L. (1957) *Microwave Measurements*, McGraw-Hill, New York.

Kim, D.I. and Naito, Y. (1982) 'Broadband design of improved hybrid-ring 3dB directional coupler' *IEEE Transactions on Microwave Theory and Techniques*, **MTT-30**, 2040–6.

March, S. (1968) 'A wideband strip line hybrid ring', *IEEE Transactions on Microwave Theory and Techniques*, **MTT-16**, 361.

Matthaei, G.L., Young, L. and Jones, E.M.T. (1965) *Microwave Filters, Impedance Matching Networks and Coupling Structures*, McGraw-Hill, New York.

Parad, L.I. and Moynihan, R.L. (1965) 'Split-tee power divider', *IEEE Transactions on Microwave Theory and Techniques*, **MTT-13**, 91–5.

Pon, C.Y. (1961) 'Hybrid ring directional coupler for arbitrary power division', *IRE Transactions on Microwave Theory and Techniques*, **MTT-9**, 529–35.

Pozar, D.M. (1990) *Microwave Engineering*, Addison-Wesley, New York.

Radmanesh, M.M. (2001) *Radio Frequency and Microwave Electronics Illustrated*, Prentice Hall, Englewood Cliffs, NJ.

Riblet, G.P. (1978) 'A directional coupler with very flat coupling', *IEEE Transactions on Microwave Theory and Techniques*, **MTT-26**, 70–4.

Rudwig, R. and Bretchko, P. (2000) *RF Circuit Design, Theory and Applications*, Prentice Hall, Englewood Cliffs, NJ.

Soares, R. (ed.) (1988) *GaAs MESFET Circuit Design*, Artech House, London.

Weber, R.J. (2001) *Introduction to Microwave Circuits: Radio Frequency and Design Applications*, IEEE Press.

Wu, Z. and Davis, L.E. (1994) 'Automation-orientated techniques for quality factor measurement of high-Tc superconducting resonators', *IEE Proceeding Part A*, **141** (6), 527–30.

5

Simulation of Microwave Devices and Circuits

ABSTRACT

In this chapter, examples of the devices described in Chapter 3 and the circuits designed using the principles or procedures described in Chapter 4 are presented. The simulation of the designed circuits using the Circuit Simulator described in Chapter 2 is also illustrated. Each circuit example is saved in the circuit holder of the same name as the circuit for easy reference, and all the examples are contained in the CD enclosed.

KEYWORDS

Microwave circuits, Transmission lines, Impedance matching, Impedance transformers, Resonators, Power dividers, Couplers, Filters, Amplifier design, Wireless transmission systems

As detailed in Chapter 3, the Software VNA has 35 built-in devices. The properties of each device can be simulated with various inputs and investigated using the display and marker functions of the Software VNA. The start and stop frequencies can be chosen to have a wide or close view of the S-parameter responses. With the design principles described in Chapter 4, a great number of microwave circuits can be designed for different applications. In this chapter, examples of the devices described in Chapter 3 and the circuits designed using the principles or procedures described in Chapter 4 are presented. The simulation of the designed circuits using the Circuit Simulator described in Chapter 2 is also illustrated. Each circuit example is saved in the circuit holder of the same name as the circuit for easy reference. All the examples in this chapter are contained in the CD and should be copied to the computer manually during the installation.

Software VNA and Microwave Network Design and Characterisation   Zhipeng Wu
© 2007 John Wiley & Sons, Ltd

5.1   TRANSMISSION LINES

5.1.1   Terminated Transmission Line

Figure 5.1(a) shows the circuit of a terminated transmission line with characteristic impedance $Z_0 = 100\,\Omega$, length $l = 0.15\,\text{m}$, phase velocity $v_\text{p} = c$ where c is the speed of light in air and load impedance $Z_\text{L} = 50\,\Omega$. The

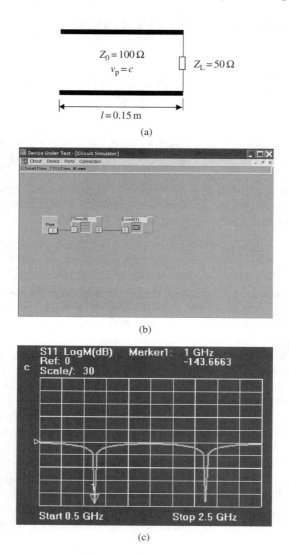

Figure 5.1   Terminated transmission line: (a) circuit diagram, (b) assembled circuit and (c) simulation results

circuit can be simulated by defining a transmission line section and a load resistance using the Circuit Simulator. The assembled circuit, Tline_ttl.nwk, is shown in Figure 5.1(b). Port 1 of the Software VNA is connected to the input port of the transmission line device, and the resistive load is connected to the output port of the transmission line. With the settings of start frequency to 0.5 GHz and stop frequency to 2.5 GHz, the simulated S_{11} response is shown in Figure 5.1(c). At 1 GHz, $|S_{11}|$ has a very small value. At this frequency, the transmission line is one half-wavelength long, and the input impedance is the same as the load impedance, i.e. 50 Ω. This repeats at 2, 3 GHz and so on when the length of the transmission line is an integer number of half-wavelengths.

5.1.2 Two-Port Transmission Line

Figure 5.2(a) shows a lossless transmission line section in two-port configuration. The transmission line has a characteristic impedance $Z_0 = 100 \, \Omega$, length $l = 0.15$ m and phase velocity $v_p = c$. The frequency responses of this transmission line section can be simulated using the General Transmission Line device described in Section 3.6. The device, Tline_2ptl.dev, is shown in Figure 5.2(b). With the settings of start frequency to 0.5 GHz and stop frequency to 2.5 GHz, the simulated S_{21} response is shown in Figure 5.2(c). At 1 GHz, the transmission line is one half-wavelength long so that $|S_{21}|$ has the largest value, i.e. $|S_{21}| = 1$ (or 0 dB). This repeats at 2, 3 GHz and so on. At 0.5, 1.5 GHz and so on, the length of the transmission line is an odd number of quarter-wavelengths so that $|S_{21}|$ has the minimum value. The S_{11} response is the same as the example in Section 5.1.1.

5.1.3 Short-Circuited Transmission Line Stub

Figure 5.3(a) shows a short-circuited transmission line stub. The transmission line stub is lossless and has a characteristic impedance $Z_0 = 50 \, \Omega$, length $l = 0.075$ m and phase velocity $v_p = c$. The frequency responses of this transmission line section can be simulated using the Two-Port Parallel Transmission Line Stub device described in Section 3.8. The device, Tline_scs.dev, is shown in Figure 5.3(b). The simulated S_{21} response is shown in Figure 5.3(c), with the start frequency of 0.5 GHz and stop frequency of 2.5 GHz. At 2 GHz, the transmission line stub is one half-wavelength long so that the input end of the stub is effectively short-circuited. The short-circuited stub has therefore a bandstop response around 2 GHz and multiples of 2 GHz.

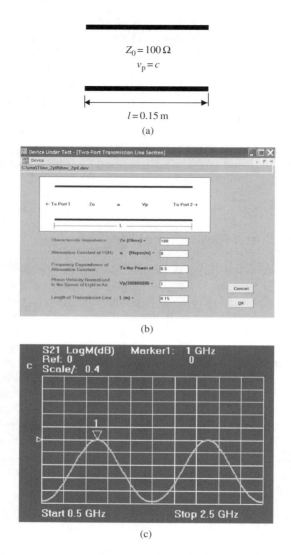

Figure 5.2 Two-port transmission line: (a) circuit diagram, (b) assembled circuit and (c) simulation results

5.1.4 Open-Circuited Transmission Line Stub

Figure 5.4(a) shows an open-circuited transmission line stub. The transmission line stub is lossless and has a characteristic impedance $Z_0 = 50\,\Omega$, length $l = 0.075\,\text{m}$ and phase velocity $v_p = c$. The frequency responses of this transmission line section can also be simulated using the Two-Port Parallel Transmission Line Stub device described in Section 3.8. The

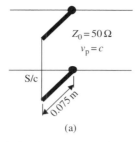

(a)

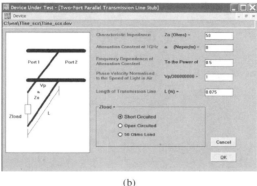

(b)

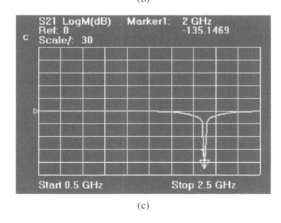

(c)

Figure 5.3   Short-circuited transmission line stub: (a) circuit diagram, (b) assembled circuit and (c) simulation results

device, Tline_ocs.dev, is shown in Figure 5.4(b). The simulated S_{21} response is shown in Figure 5.4(c), with the start frequency of 0.5 GHz and stop frequency of 2.5 GHz. At 1 GHz, the transmission line stub is one quarter-wavelength long so that the input end of the stub is effectively short-circuited. The open-circuited stub has therefore a bandstop response around 1 GHz and odd multiples of 1 GHz.

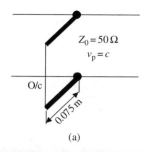

(a)

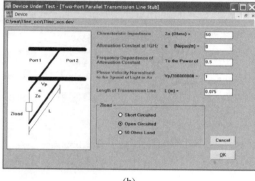

(b)

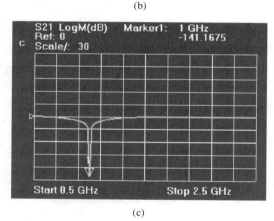

(c)

Figure 5.4   Open-circuited transmission line stub: (a) circuit diagram, (b) assembled circuit and (c) simulation results

5.1.5   Periodic Transmission Line Structures

Figure 5.5(a) shows a periodic structure made of microstrip line sections and open-circuited stubs. The purpose of this example is to illustrate the filtering response of the structure. The microstrip lines and stubs are made of RT/Duroid 5880 with the following parameters: $\varepsilon_r = 2.2$, $w = 2.575$ mm,

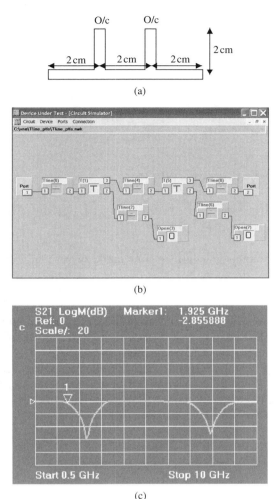

Figure 5.5   Periodic transmission line structure: (a) circuit diagram, (b) assembled circuit and (c) simulation results

$h = 0.79$ mm, $L = 2$ cm. The width of the line, w, is chosen using the Microstrip Line device described in Section 3.15 so that $Z_0 = 50\,\Omega$.

The periodic structure can be simulated by defining the transmission line sections and stubs, which are connected by T-junctions. The circuit assembled using the Circuit Simulator, Tline_ptls.nwk, is shown in Figure 5.5(b). Port 1 of the Software VNA is connected to the input port of the first transmission line section and Port 2 to the last transmission line section. The start frequency is set to 0.5 GHz and stop frequency to

10 GHz. The simulated S_{21} response is shown in Figure 5.5(c). The periodic structure has a low-pass response at 1.9 GHz, bandstop response at 2.7 GHz when the stub is a quarter-wavelength long, bandpass response at 5.4 GHz (with 3 dB bandwidth of 3.6 GHz) when the stub is one half-wavelength long.

The periodic structure can be modified with short-circuited stubs. The frequency responses of the modified periodic structure, Tline_ptls2.nwk, can be investigated further.

5.2   IMPEDANCE MATCHING

5.2.1   Matching of a Half-Wavelength Dipole Antenna Using a Discrete Element

For a dipole antenna of length 50 mm made of a metallic wire of radius $a = 0.5$ mm, the dipole is one half-wavelength long at $f_0 = 3$ GHz. The input impedance of the dipole at 3 GHz can be obtained using the Dipole Antenna device described in Section 3.30, with the use of Smith chart display and frequency span from 2 to 4 GHz. The dipole antenna is shown in Figure 5.6(a). The input impedance and admittance of the antenna at 3 GHz are $Z_{ant}(f_0) = 73.13 + j42.54\,\Omega$ and $Y_{ant}(f_0) = G_{ant}(f_0) + jB_{ant}(f_0) = 0.010217 - j0.005943$ S, respectively. The reflection coefficient at 3 GHz is $|S_{11}(f_0)| = -8.6$ dB without matching. The dipole antenna can be matched using a capacitor connected in parallel with the antenna and a quarter-wave transformer. The required capacitance value is

$$C = -\frac{B_{ant}(f_0)}{2\pi f_0} = 3.153 \times 10^{-13}\,F = 0.3153\,pF$$

and the characteristic impedance of the transformer is

$$Z_0 = \sqrt{50 G_{ant}^{-1}(f_0)} = 69.96\,\Omega.$$

The length of the transformer is 25 mm with $v_p = c$, which is one quarter-wavelength at 3 GHz. The matching circuit can be assembled using the Circuit Simulator with a dipole antenna, a capacitor, a transmission line section and the parameters calculated above. The input end of the transmission line section is connected to Port 1 of the Software VNA. The circuit, Matching_de.nwk, is shown in Figure 5.6(b). The simulated results are shown in Figure 5.6(c) with the frequency span from 1 to 5 GHz. The dipole is well matched at 3 GHz with -10 dB bandwidth of 17.8% from 2.75 to 3.24 GHz.

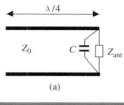

(a)

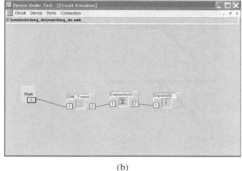

(b)

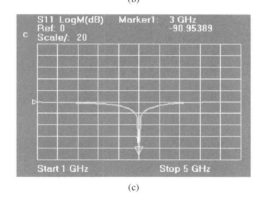

(c)

Figure 5.6 Matching of a dipole antenna using a discrete element: (a) circuit diagram, (b) assembled circuit and (c) simulation results

5.2.2   Single Stub Matching of a Half-Wavelength Dipole Antenna

The half-wavelength dipole of length 50 mm and wire radius 0.5 mm shown in Figure 5.6(a) can also be matched using the single stub matching technique described in Section 4.1.2. The input impedance of the dipole at $f_0 = 3$ GHz is $Z_{ant}(f_0) = 73.13 + j42.54\,\Omega$. Using the equations in Section 4.1.2, the required lengths of the transmission line section and short-circuited stub with $Z_0 = 50\,\Omega$ can be calculated to be $l_1 = 0.02142$ m and $l_2 = 0.014254$ m, respectively. The matching circuit can be assembled using the Circuit Simulator with a dipole antenna, a 50 Ω transmission line section, a $50\,\Omega$ stub and the specified lengths above. The input end of the transmission line stub is connected to Port 1 of the Software VNA. The circuit, Matching_ss.nwk, is shown in Figure 5.7(a). The simulated results are shown in Figure 5.7(b) with the frequency span from 1 to 5 GHz. The dipole is well matched at 3 GHz with $-10\,dB$ bandwidth of 13.9% from 2.8 to 3.19 GHz.

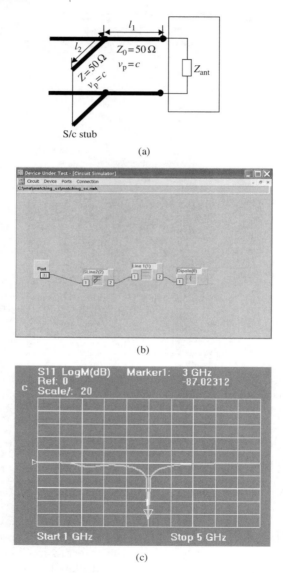

(a)

(b)

(c)

Figure 5.7 Single stub matching of a dipole antenna: (a) circuit diagram, (b) assembled circuit and (c) simulation results

5.3   IMPEDANCE TRANSFORMERS

5.3.1   Quarter-Wave Impedance Transformer

Figure 5.8(a) shows the design of a quarter-wave impedance transformer which transforms the load resistance $R_L = 10 - 50\,\Omega$ input impedance

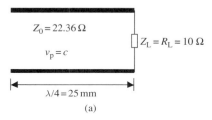

(a)

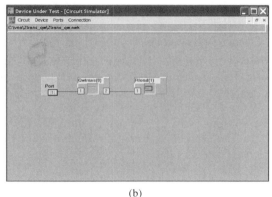

(b)

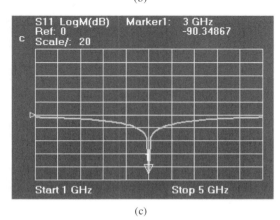

(c)

Figure 5.8 Quarter-wave impedance transformer: (a) circuit diagram, (b) assembled circuit and (c) simulation results

at 3 GHz. The length of transmission line with $v_p = c$ is 25 mm. The characteristic impedance of the line can be calculated using the equation in Section 4.2.1, i.e.

$$Z_0 = \sqrt{50 \times 10} = 22.36 \, \Omega.$$

The theoretical bandwidth of the transformer at VSWR = 1.5 is 29%.

Using the Circuit Simulator, the assembled quarter-wave transformer circuit, Ztrans_qw.nwk, is as shown in Figure 5.8(b). It consists of a transmission line section and a resistive load with the parameters specified above. The input end of the transformer is connected to Port 1 of the Software VNA. The simulated results are shown in Figure 5.8(c) with the frequency span from 1 to 5 GHz. Using the VSWR display, it can be determined that the quarter-wave transformer has a bandwidth of 29.3% at VSWR = 1.5.

5.3.2   Chebyshev Multisection Impedance Transformer

Figure 5.9(a) shows the design of a four-section Chebyshev impedance transformer using the equations in Section 4.2.2 with the maximum reflection in the matched band $\Gamma_m = 0.2$, the central frequency $f_0 = 3$ GHz, load resistance $R_L = 100 \, \Omega$ and source impedance $Z_s = Z_0 = 50 \, \Omega$. All transmission line sections with $v_p = c$ are 25 mm in length, which is $\lambda/4$ long at 3 GHz. The required characteristic impedances of the line sections are calculated to be (Pozar, 1990)

$$Z_{01} = 63.13 \, \Omega, \quad Z_{02} = 67.48 \, \Omega, \quad Z_{03} = 72.31 \, \Omega \quad \text{and} \quad Z_{04} = 77.30 \, \Omega.$$

Theoretically, for $N = 4$, $\Gamma_m = 0.2$, $R_L = 100 \, \Omega$ and $Z_0 = 50 \, \Omega$,

$$\sec \theta_m = \cosh \left[\frac{1}{4} \cosh^{-1} \left(\frac{1}{0.2} \left| \frac{100 - 50}{100 + 50} \right| \right) \right] = \cosh \left[\frac{1}{4} \cosh^{-1} \left(\frac{5}{3} \right) \right]$$

so that $\theta_m = 15.54° = 0.27126$ rad and the bandwidth is

$$\frac{\Delta f}{f_0} \Big|_{\text{at VSWR}=1.5 \text{ or} \Gamma_m = 0.2} = 2 - \frac{4\theta_m}{\pi} = 1.655 = 165.5\%.$$

Using the Circuit Simulator, the assembled quarter-wave transformer circuit, Ztrans_Cheby.nwk, is as shown in Figure 5.9(b). The simulated results are shown in Figure 5.9(c) with the frequency span from 0 to 6 GHz. Chebyshev ripples can be observed in the display. Using the VSWR format, it can be determined that the four-section impedance transformer has a bandwidth of 163% at VSWR = 1.5 from 0.6 to 5.5 GHz. The maximum

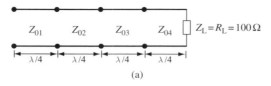

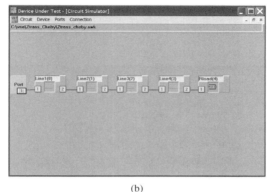

Figure 5.9 Chebyshev multisection impedance transformer: (a) circuit diagram, (b) assembled circuit and (c) simulation results

reflection in the matched band is $\Gamma_m = 0.21$ with the assigned transmission line parameters, which can be seen using the linear display, LinM.

5.3.3 Corporate Feeds

Figure 5.10(a) shows two corporate feeds that are commonly used in the impedance matching of antenna arrays. In the first design, two 50 Ω lines are connected in parallel. The resultant impedance is transformed to 50 Ω using a 35.36 Ω quarter-wave transformer. In the second design, two 70.7 Ω quarter-wave transformers are used. Each transformer transforms

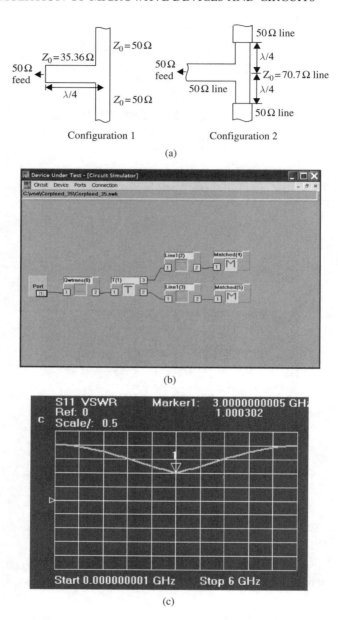

Figure 5.10   Corporate feeds: (a) circuit diagrams, (b) assembled circuit of configuration 1 and (c) simulation results of configuration 1

the $50\,\Omega$ load impedance to an impedance of $100\,\Omega$ at the central frequency. The transformed impedances are connected in parallel to give a resultant impedance of $50\,\Omega$. In both designs, the length of each transformer is 25 mm for a central frequency of 3 GHz when a transmission line with $v_p = c$ is used.

The assembled circuit for the first design, Corpfeed_35.nwk, is shown in Figure 5.10(b). The 50 Ω load impedance is represented by a Matched Load. The simulated results with a frequency span from 0 to 6 GHz are shown in Figure 5.10(c). The impedance is very well matched at 3 GHz. Using the VSWR display, it can be seen that the VSWR values of both corporate feeds are always less than 2. The assembled circuit for the second design, Corpfeed_70.nwk, can also be simulated in the same way.

5.3.4 Corporate Feeds Realised Using Microstrip Lines

The corporate feeds discussed in Section 5.3.4 can be realised using microstrip lines. Figure 5.11(a) shows the microstrip version of the design in Figure 5.10(a), realised using RT/Duroid 5880 with $\varepsilon_r = 2.2$ and substrate thickness $h = 0.79$ mm. By setting the start frequency to 3 GHz and using the Microstrip Line device described in Section 3.15, the width and the length of the 50 Ω microstrip line can be determined or chosen to be

$$w_1 = 2.48\,\text{mm}, \quad L = \frac{\lambda}{4} = 0.018161\,\text{m for } \varepsilon_{re} = 1.895.$$

Similarly the width and the length of the 35.36 Ω microstrip line can be determined to be

$$w_2 = 4.14\,\text{mm}, \quad L = \frac{\lambda}{4} = 0.017839\,\text{m for } \varepsilon_{re} = 1.964.$$

The assembled circuit, Corpfeed_35m.nwk, is shown in Figure 5.11(b). The 50 Ω load impedance is again represented by a Matched Load. The simulated results with a frequency span from 0 to 6 GHz are shown in Figure 5.11(c). The impedance is also very well matched at 3 GHz.

5.3.5 Kuroda's Identities

Kuroda's identities are often used in microwave filter design using transmission lines (Pozar, 1990). Figure 5.12(a) and (b) shows one of Kuroda identities. It shows that an open-circuited stub of characteristic impedance Z_2 and length $\lambda/8$ connected with a transmission line section of characteristic impedance Z_1 and length $\lambda/8$ is equivalent to a transmission line section of characteristic impedance Z_2/n^2 and length $\lambda/8$ connected with a short-circuited stub of characteristic impedance Z_1/n^2 and length $\lambda/8$ where

$$n^2 = 1 + Z_2/Z_1.$$

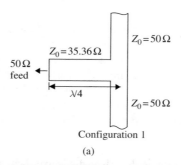

Configuration 1

(a)

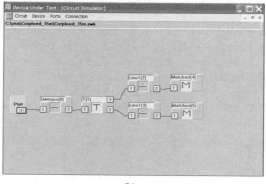

(b)

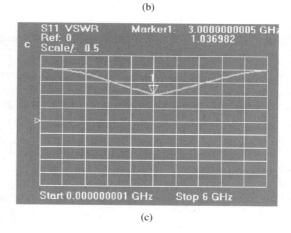

(c)

Figure 5.11 Corporate feed using microstrip lines: (a) circuit diagram, (b) assembled circuit and (c) simulation results

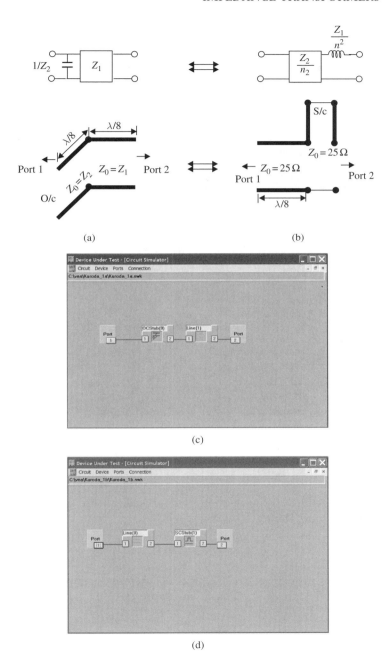

Figure 5.12 Kuroda's identities: (a) and (b) identity circuits, (c) and (d) assembled circuits and (e) and (f) simulation results

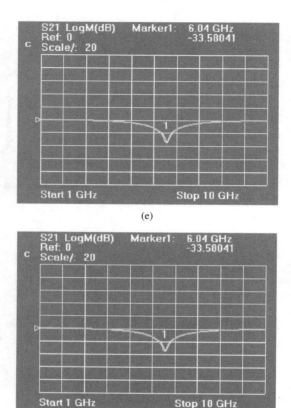

(e)

(f)

Figure 5.12   (Continued)

When a transmission line with $v_p = c$ is used, all line sections have a length of 12.5 mm. Taking

$$Z_1 = Z_2 = 50 \, \Omega$$

for the circuit in Figure 5.12(a), the characteristic impedance of the line section and stub of the circuit in Figure 5.12(b) can be calculated to be

$$\frac{Z_2}{n^2} = 25 \, \Omega$$

with $n^2 = 2$. The assembled circuits for Figure 5.12(a) and (b), i.e. Kuroda_1a.nwk and Kuroda_1b.nwk, are shown in Figure 5.12(c) and (d), respectively. The simulation results with the frequency span from 1 to 10 GHz of these two circuits are shown in Figure 5.12(e) and (f), respectively. The identity of both circuits can be examined in terms of all S-parameters and also in terms of amplitude and phase.

5.4 RESONATORS

5.4.1 One-Port RLC Series Resonant Circuit

Figure 5.13(a) shows a one-port directly connected RLC series resonant circuit with

$$R = 10\,\Omega, \quad L = 30\,\text{nH}, \quad C = 0.03\,\text{pF}.$$

The resonance frequency of the resonant circuit is

$$f_0 = \frac{1}{2\pi\sqrt{LC}} = 5.3\,\text{GHz}$$

and the unloaded and loaded Q-factors are

$$Q_0 = \frac{\omega_0 L}{R} = \frac{\sqrt{L/C}}{R} = 100$$

and

$$Q_L = \frac{\omega_0 L}{R + Z_0} = \frac{1000}{10 + 50} = 16.7,$$

respectively. The resonant circuit can be simulated using the One-Port Impedance Load device described in Section 3.3 as Reson_1psr.dev, with the frequency span from 5.1 to 5.5 GHz. The results of the $|S_{11}|$ response are shown in Figure 5.13(b). Use Marker 1 to locate the resonance frequency and the Q_0 and Q_L measurement functions to measure their values. The Q-factors can be more accurately measured with a smaller frequency span or an increased number of frequency points.

5.4.2 Two-Port RLC Series Resonant Circuit

Figure 5.14(a) shows the same RLC series resonant circuit as that in Section 5.4.1 but in two-port connections. The resonance frequency and unloaded Q-factor remain the same. But the loaded Q-factor is changed to

$$Q_L = \frac{\omega_0 L}{R + 2Z_0} = \frac{1000}{110} = 9.09.$$

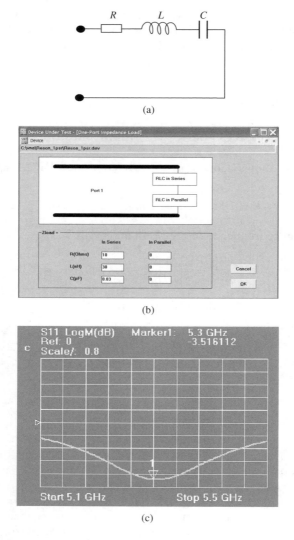

Figure 5.13 One-port RLC series resonant circuit: (a) circuit diagram, (b) assembled circuit and (c) simulation results

The resonant circuit can be simulated using the Two-Port Single Impedance device described in Section 3.4 as Reson_2psr.dev, with the frequency span from 5.1 to 5.5 GHz. The results of the $|S_{21}|$ response are shown in Figure 5.14(b). Use Marker 1 to locate the resonance frequency and the Q_0 and Q_L measurement functions to measure their values. The Q-factors can be measured using either S_{21} or S_{11} display. The resonance

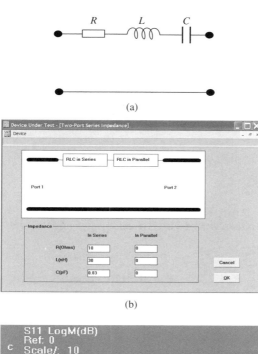

(a)

(b)

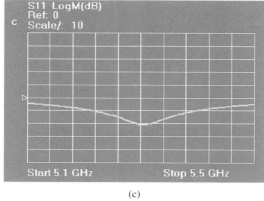

(c)

Figure 5.14 Two-port RLC series resonant circuit: (a) circuit diagram, (b) assembled circuit and (c) simulation results

response of S_{21} can also be viewed in the two-port chant, 2PChart display, described in Section 1.4.3.

The resonance responses can be further investigated with the following parameters:

$$R = 500\,\Omega, \quad L = 3\,\mu H, \quad C = 3 \times 10^{-4}\,pF$$

as in Reson_2psr2dev. The resonant circuit has the same resonance frequency, but a higher Q-factor with

$$Q_0 = 200 \quad \text{and} \quad Q_L = 166.7.$$

The Q-factors can be more accurately measured with a smaller frequency span.

5.4.3   Two-Port Coupled Resonant Circuit

Figure 5.15(a) shows the circuit that is commonly used to represent a microwave resonator with two-port coupling. The specific values of R, L and C are

$$R = 2\,\Omega, \quad L = 30\,\text{nH}, \quad C = 0.03\,\text{pF}.$$

The coupling coefficient is $N = 0.1$. The resonant circuit has a resonance frequency $f_0 = 5.3\,\text{GHz}$ and unloaded Q-factor $Q_0 = 500$.

The assembled circuit, Reson_2pcr.nwk, using the Circuit Simulator is shown in Figure 5.15(b). The results of the $|S_{21}|$ response are shown in Figure 5.15(c) with the frequency span from 5.2 to 5.4 GHz. Use Marker 1 to locate the resonance frequency and the Q_0 and Q_L measurement functions to measure their values. The Q-factors can be measured using either S_{21} or S_{11} display. The resonance response of S_{21} can also be viewed in the two-port chart display, 2PChart, described in Section 1.4.3. The effect of coupling coefficient to the S_{21} response can be investigated by changing the value of N.

5.4.4   Two-Port Coupled Microstrip Line Resonator

Figure 5.16(a) shows a two-port end-coupled microstrip resonator. The resonator is formed of two coupling gaps and a microstrip line. The microstrip line is made of RT/Duroid 5880 with $\varepsilon_r = 2.2$, thickness $h = 0.79\,\text{mm}$, width $w = 2.48\,\text{mm}$ and length $L = 30\,\text{mm}$. The coupling gap is $s = 1\,\text{mm}$. The effective dielectric constant of the microstrip line is $\varepsilon_{re} = 1.895$ and the characteristic impedance of the line is approximately $50\,\Omega$. The first resonance occurs at f_{01} when the microstrip line is $\lambda/2$ long. Higher order resonance at $f_{0m} = m\,f_{01}$ is $m\lambda/2$ where m is an integer.

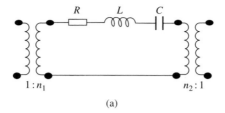

(a)

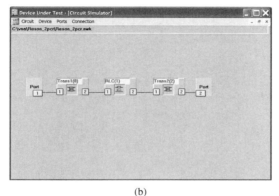

(b)

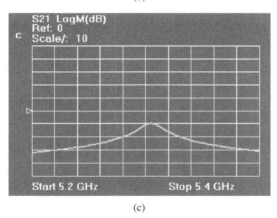

(c)

Figure 5.15 Two-port coupled RLC series resonant circuit: (a) circuit diagram, (b) assembled circuit and (c) simulation results

The assembled circuit, Reson_2pm.nwk, using the Circuit Simulator is shown in Figure 5.16(b). The results of the $|S_{21}|$ response are shown in Figure 5.16(c) with the frequency span from 3.4 to 3.5 GHz. Use Marker 1 to locate the resonance frequency at 3.465 GHz and the Q_0 and Q_L measurement functions to measure their values. The Q_0 and Q_L values are 190 and 159, respectively, with the inclusion of conducting and radiation losses.

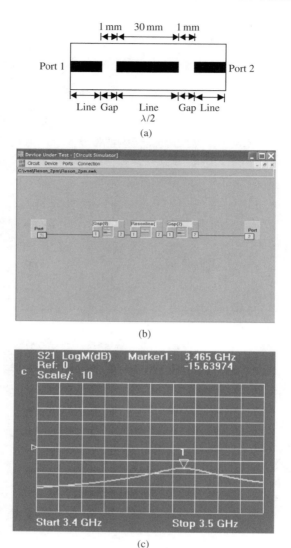

Figure 5.16 Two-port coupled microstrip line resonator: (a) resonator layout, (b) assembled circuit and (c) simulation results

5.4.5 Two-Port Coupled Microstrip Coupled Line Resonator

Figure 5.17(a) shows a two-port coupled microstrip coupled line resonator. The resonator is formed of two coupling gaps and a microstrip coupled

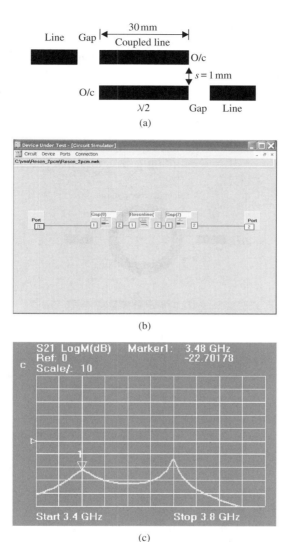

Figure 5.17 Two-port coupled microstrip coupled line resonator: (a) resonator layout, (b) assembled circuit and (c) simulation results

line. The coupled line is made of RT/Duroid 5880 with $\varepsilon_r = 2.2$, thickness $h = 0.79$ mm, width $w = 2.48$ mm and length $L = 30$ mm. The coupling gap is $s = 1$ mm. The first even and odd mode resonances occur when the length of the coupled line is $\lambda_e/2$ or $\lambda_o/2$.

The assembled circuit, Reson_2pcm.nwk, using the Circuit Simulator is shown in Figure 5.17(b). The results of the $|S_{21}|$ response are shown in Figure 5.17(c) with the frequency span from 3.4 to 3.8 GHz. Use Marker 1 to locate the resonance frequencies of the even mode at 3.48 GHz and odd mode at 3.64 GHz. Use the Q_0 and Q_L measurement functions to measure their values. The Q_0 value of the even mode is lower than that of the odd mode due to a higher loss of the even mode.

5.4.6 Two-Port Symmetrically Coupled Ring Resonator

Figure 5.18(a) shows a two-port symmetrically coupled ring resonator. The resonator is formed of two coupling gaps and a microstrip ring. The ring

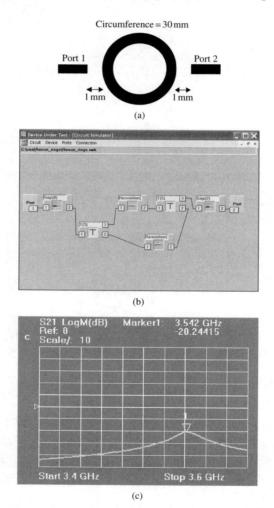

Figure 5.18 Two-port symmetrically coupled ring resonator: (a) resonator layout, (b) assembled circuit and (c) simulation results

is made of RT/Duroid 5880 with $\varepsilon_r = 2.2$, thickness $h = 0.79$ mm, width $w = 2.48$ mm and circumference 60 mm. The coupling gap is $s = 1$ mm. The effective dielectric constant of the microstrip line is $\varepsilon_{re} = 1.895$ and the characteristic impedance of the line is approximately $50\,\Omega$. The first resonance frequency is estimated to be at 3.47 GHz when the circumference of the ring is one wavelength.

The assembled circuit, Reson_rings.nwk, using the Circuit Simulator is shown in Figure 5.18(b). The ring is represented by two transmission lines connected in parallel, and two T-junctions are used to connect the transmission lines and the coupling gaps. The results of the $|S_{21}|$ response are shown in Figure 5.18(c) with the frequency span from 3.4 to 3.6 GHz. Use Marker 1 to locate the resonance frequency at 3.542 GHz, which is different from the theoretical value due to the effect of the coupling gaps. Using the Q-factor measurement functions, the Q_0 and Q_L factors are measured to be 196 and 177, respectively, with the inclusion of conducting and radiation losses.

5.4.7   Two-Port Asymmetrically Coupled Ring Resonator

Figure 5.19(a) shows a two-port asymmetrically coupled ring resonator. The resonator is made of RT/Duroid 5880 with $\varepsilon_r = 2.2$, thickness $h = 0.79$ mm, width $w = 2.48$ mm and circumference 60 mm. However, the ring is coupled asymmetrically with an arc length of 31.5 mm at the top and 28.5 mm at the bottom. The resonator is coupled by two transformers with $N = 0.01$.

The assembled circuit, Reson_ringas.nwk, using the Circuit Simulator is shown in Figure 5.19(b). The ring is represented by two transmission lines of different lengths connected in parallel, and two T-junctions are used to connect the transmission lines and the transformers. The results of the $|S_{21}|$ response are shown in Figure 5.19(c) with the frequency span from 3.3 to 3.9 GHz. Due to the asymmetrical coupling, there exist two closely spaced resonances in the frequency range. Use Marker 1 to locate the resonance frequencies at 3.456 and 3.816 GHz. The Q_0 and Q_L factors of the resonances can also be measured with a smaller frequency span using the Q-factor measurement functions.

5.5   POWER DIVIDERS

5.5.1   3 dB Wilkinson Power Divider

Figure 5.20(a) shows the design of a 3 dB Wilkinson power divider described in Section 4.4.1 with a central frequency $f_0 = 3$ GHz. For the port impedance

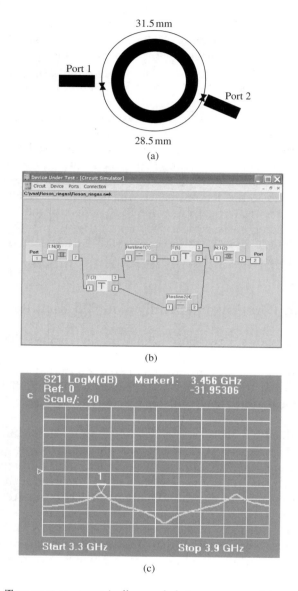

Figure 5.19   Two-port asymmetrically coupled ring resonator: (a) resonator layout, (b) assembled circuit and (c) simulation results

of $Z_0 = 50\,\Omega$, the characteristic impedance of the quarter-wave transmission line sections is $\sqrt{2}Z_0 = 70.7\,\Omega$, and the length of the line is 25 mm assuming $v_p = c$. The resistance of the isolation resistor connecting the two arms is $R = 100\,\Omega$.

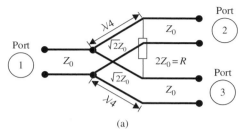

(a)

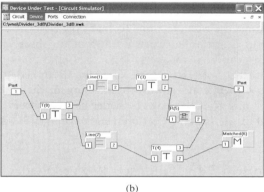

(b)

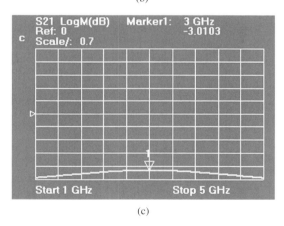

(c)

Figure 5.20 3 dB Wilkinson power divider: (a) circuit diagram, (b) assembled circuit and (c) simulation results

The Wilkinson power divider can be assembled using the Circuit Simulator as shown in Figure 5.20(b). The circuit, Divider_3dB.nwk, consists of two transmission lines, a series resistor in two-port connection, three T-junctions and a Matched Load. The Matched Load is used to terminate Port 3 of the power divider. Port 2 of the power divider serves as the measurement port, which is connected to Port 2 of the Software VNA.

The S_{21} response is shown in Figure 5.20(c) with the frequency span from 1 to 5 GHz. The S_{21} value at 3 GHz is -3.01 dB, which can be identified using Marker 1. The 0.2 dB coupling deviation bandwidth is 82.7% from 1.76 to 4.24 GHz. Using the Phase display, it can be seen that there is a phase delay of $90°$ at 3 GHz at the output port, which agrees with the theoretical value in Section 4.4.1.

When the VSWR of the reflection response, S_{11}, is displayed, it can be measured that the bandwidth at VSWR $= 1.22$ is 37% from 2.44 to 3.56 GHz.

It is possible to disconnect Port 2 of the Software VNA and the Matched Load and exchange their connections so that Port 3 of the power divider is connected to Port 2 of the Software VNA. Similar observations of the S_{21} and S_{11} responses can be made.

If Port 1 of the Software VNA and the Matched Load are both disconnected and their connections are exchanged, the isolation response between Ports 2 and 3 of the power divider can be observed from the S_{21} display. At 3 GHz, the isolation between the two ports is 82.4 dB.

5.5.2   Microstrip 3 dB Wilkinson Power Divider

The 3 dB Wilkinson power divider in Section 5.5.2 can be realised using microstrip lines. Figure 5.21(a) shows the microstrip version of the 3 dB Wilkinson power divider, made of RT/Duroid 5880 with $\varepsilon_r = 2.2$ and thickness $h = 0.79$ mm. Using the Microstrip Line device described in Section 3.15 and setting the start frequency to 3 GHz, the width of the required $70.7\,\Omega$ line section at 3 GHz can be found to be $w = 1.392$ mm, and the effective relative dielectric constant at 3 GHz is $\varepsilon_{re} = 1.823$. Hence the length of the line section is

$$L = \frac{\lambda}{4} = \frac{0.025}{\sqrt{\varepsilon_{re}}} = 0.018516 \, \text{m}.$$

With these parameters, the circuit of the microstrip 3 dB Wilkinson power divider can be assembled and the radiation losses of the microstrip lines can be included in the modelling. The circuit, Divider_3dBm.nwk, is shown in Figure 5.21(b) with Port 2 of the power divider connected to Port 2 of the Software VNA. The S_{21} response of the power divider is shown in Figure 5.21(c) with the frequency span from 1 to 5 GHz. The S_{21} value at 3 GHz is -3.04 dB. The 0.2 dB deviation bandwidth is 82.7% from 1.68 to 4.16 GHz. Using the Phase display, it can be seen that there is a phase delay of $90.01°$ at 3 GHz at the output port, which agrees well with the theoretical value. The VSWR and isolation responses can be observed in the same way as those in Section 5.5.1.

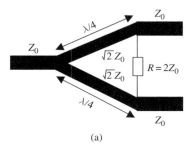

(a)

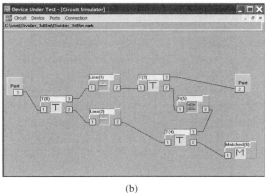

(b)

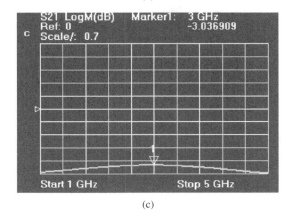

(c)

Figure 5.21 Microstrip 3 dB Wilkinson power divider: (a) circuit diagram, (b) assembled circuit and (c) simulation results

5.5.3 Cohn's Cascaded 3 dB Power Divider

Figure 5.22(a) shows the design of Cohn's cascaded power divider with 3 dB division at $f_0 = 3$ GHz using the equations in Section 4.4.4. For $N = 2$

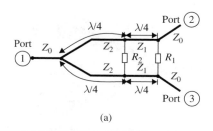

(a)

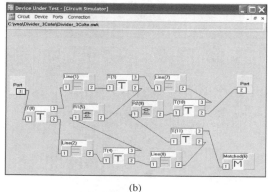

(b)

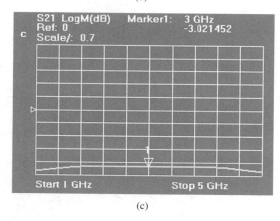

(c)

Figure 5.22   Cohn's cascaded 3 dB power divider: (a) circuit diagram, (b) assembled circuit and (c) simulation results

and the frequency bandwidth ratio of $f_2/f_1 = 2$, the required Z_1, Z_2, R_1 and R_2 values are (Cohn, 1968)

$$Z_1 = 1.2197 \times 50 = 60.99\,\Omega; \quad Z_2 = 1.6398 \times 50 = 81.99\,\Omega;$$
$$R_1 = 4.8204 \times 50 = 241\,\Omega; \quad R_2 = 1.9602 \times 50 = 98\,\Omega.$$

The length of the quarter-wavelength transmission line sections with $v_p = c$ is $\lambda/4 = 0.025\,m$. With these parameters, the circuit of the power divider can be assembled, as shown in Figure 5.22(b). The circuit, Divider_3Cohn.nwk, consists of four transmission line sections, two isolation resistors, five T-junctions and a Matched Load. With Port 2 of the Software VNA connected to Port 2 of the divider, the simulated results of the S_{21} response are shown in Figure 5.22(c) with the frequency span from 1 to 5 GHz. The value at 3 GHz is 3.02 dB. The 0.2 dB deviation bandwidth is 112.7% from 1.16 to 4.84 GHz, which is wider than the Wilkinson power divider. Using the Phase display, it can be seen that the phase delay at the output port at 3 GHz is 180°.

When the VSWR of the reflection response, S_{11}, is displayed, it can be measured that the bandwidth at VSWR = 1.22 is 84% from 1.76 to 4.28 GHz.

When Ports 1 and 2 of the Software VNA are connected to Ports 2 and 3 of the power divider, as in Divider_3Cohn2.nwk, the isolation response can be observed from the S_{21} display. The isolation at 3 GHz is 27.3 dB, which is worse than the Wilkinson power divider.

5.6 COUPLERS

5.6.1 Two-Stub Branch Line Coupler

Figure 5.23(a) shows the design of a 3 dB 90° two-stub branch line coupler with the central frequency of $f_0 = 3\,GHz$. The characteristic impedances of the line sections are $Z_a = 35.4\,\Omega$ and $Z_b = 50\,\Omega$. The length of the line sections with $v_p = c$ is 25 mm. The circuit of the branch line coupler assembled using the Circuit Simulator, Coupler_2sbl.nwk, is shown in Figure 5.23(b) with the frequency span from 1 to 5 GHz. It consists of four line sections, four T-junctions and two Matched Loads. Port 2 of the Software VNA is connected to Port 3 of the coupler. The results of the coupling response, $|S_{21}|$, are shown in Figure 5.23(c). The coupling strength at 3 GHz is −3 dB and the phase delay is 180°. The 0.2 dB deviation bandwidth is 33.3% from 2.5 to 3.5 GHz. The return loss at 3 GHz is over 57 dB at both ports.

If the connections of Port 2 of the Software VNA and the Matched Load at Port 2 of the coupler are exchanged, the direct transmission response can be observed. The phase delay at 3 GHz is 90°. Similarly if Port 2 of the Software VNA is connected to Port 4 of the coupler, the isolation characteristics can be observed.

For a design of 10 dB 90° two-stub branch line coupler with the central frequency of $f_0 = 3\,GHz$, the characteristic impedances of the line sections

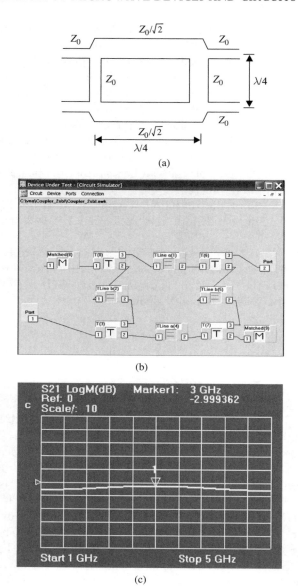

Figure 5.23 Two-stub branch line coupler: (a) circuit diagram, (b) assembled circuit and (c) simulation results

are required to be $Z_a = 47.4\,\Omega$ and $Z_b = 150\,\Omega$. The coupling, direct transmission and isolation responses of the coupler can be simulated using the circuit assembled in Coupler_2sbl2.nwk.

5.6.2 Microstrip Two-Stub Branch Line Coupler

The 3 dB 90° two-stub branch line coupler in Section 5.6.1 can be realised using the microstrip line made of RT/Duroid 5880 with $\varepsilon_r = 2.2$ and thickness $h = 0.79$ mm. Using the Microstrip Line device described in Section 3.15 and setting the start frequency to 3 GHz, the width of the required 35.4 Ω line section at 3 GHz can be found to be $w_a = 4.14$ mm, and the effective relative dielectric constant at 3 GHz is $\varepsilon_{re} = 1.964$. The length of the 35.4 Ω line section is thus $L_a = \lambda/4 = 0.01784$ m. Similarly for the 50 Ω line section, $w_b = 2.48$ mm, $\varepsilon_{re} = 1.895$ and $L_b = \lambda/4 = 0.01816$ m. With these parameters, the circuit of the microstrip 3 dB 90° two-stub branch line coupler can be assembled. The radiation losses of the microstrip lines can be included in the modelling. The circuit, Coupler_2sblm.nwk, is shown in Figure 5.24(a) with Port 2 of Software VNA connected to Port 3 of the coupler. The S_{21} response of the power divider is shown in Figure 5.24(b) with the frequency span from 1 to 5 GHz. The S_{21} value at 3 GHz is -3.09 dB. The direct transmission or isolation response can also be observed when Port 2 of the Software VNA is connected to Port 2 or Port 4 of the coupler, respectively.

5.6.3 Three-Stub Branch Line Coupler

Figure 5.25(a) shows the design of a 0 dB 90° three-stub branch line coupler with the central frequency of $f_0 = 3$ GHz. The characteristic impedances of the line sections are $Z_a = 50\,\Omega$, $Z_b = 50\,\Omega$ and $Z_d = 50\,\Omega$. The length of the line sections with $v_p = c$ is 25 mm. The circuit of the branch line coupler assembled using the Circuit Simulator, Coupler_3sbl.nwk, is shown in Figure 5.25(b) with the frequency span from 1 to 5 GHz. It consists of seven line sections, six T-junctions and two Matched Loads. Port 2 of the Software VNA is connected to Port 3 of the coupler. The results of the coupling response, $|S_{21}|$, are shown in Figure 5.25(c). The coupling strength at 3 GHz is 0 dB, and the phase at 3 GHz is 90° or a phase delay of 270°. The 0.2 dB deviation bandwidth is 9% from 2.5 to 3.5 GHz. The coupler is very well matched at Port 1.

For a design of 3 dB 90° three-stub branch line coupler with the central frequency of $f_0 = 3$ GHz, the characteristic impedances of the line sections are required to be $Z_a = 35.4\,\Omega$, $Z_b = 120.7\,\Omega$, $Z_d = 35.4\,\Omega$. The coupling, direct transmission and isolation responses of the coupler can be simulated using the circuit assembled in Coupler_3sbl2.nwk.

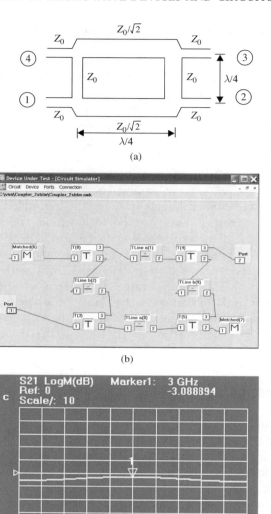

Figure 5.24 Microstrip two-stub branch line coupler: (a) circuit diagram, (b) assembled circuit and (c) simulation results

5.6.4   Coupled Line Coupler

The coupled line coupler can be simulated using the Four-Port Coupled Lines device described in Section 3.22. For a 10 dB coupled line coupler with a central frequency of $f_0 = 1.5\,\text{GHz}$, the even and odd mode characteristic

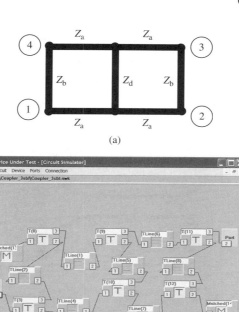

(a)

(b)

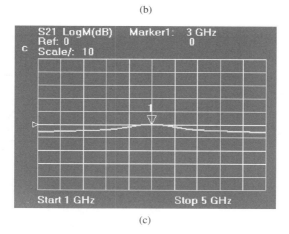

(c)

Figure 5.25 Three-stub branch line coupler: (a) circuit diagram, (b) assembled circuit and (c) simulation results

impedances of the coupled line can be obtained, using the equations in Section 4.5.4, to be $Z_{0e} = 69.4\,\Omega$ and $Z_{0o} = 36\,\Omega$, and the length of the line is $L = 50\,\text{mm}$ for $v_p = c$. Figure 5.26(a) shows the simulation input of the coupler, Coupler_cl.dev, with Port 1 of the Software VNA connected to Port 1 of the coupler and Port 2 of the Software VNA to Port 3 of the coupler. The coupling response is shown in Figure 5.26(b) with the frequency span

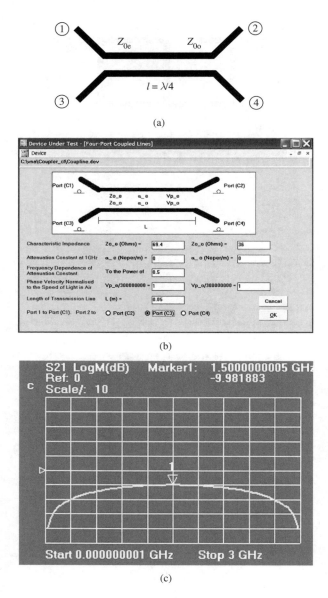

Figure 5.26 Coupled line coupler: (a) circuit diagram, (b) assembled circuit and (c) simulation results

from 0 to 3 GHz. The coupling strength at 1.5 GHz is −9.98 dB, as indicated by Marker 1. The 0.2 dB deviation bandwidth is 29%. Port 1 of the coupler is very well matched. If Port 2 of the Software VNA is connected to Port 2 or Port 4 of the coupler, the direct transmission and isolation responses of the coupler can be observed.

5.6.5 Microstrip Coupled Line Coupler

The 10 dB coupled line coupler can be realised using microstrip lines. This can be simulated using the Four-Port Coupled Microstrip Lines device described in Section 3.24. The device, Coupler_clm.dev, is shown in Figure 5.27(a). With the choice of RT/Duroid of $\varepsilon_r = 2.2$ and thickness $h = 0.79$ mm, the width of the microstrip lines and the spacing between the lines are determined to be $w = 2.025$ mm and $s = 0.15$ mm for $Z_{0e} = 69.529\,\Omega$ and $Z_{0o} = 35.978\,\Omega$. At $f_0 = 1.5$ GHz, $\varepsilon_{ree} = 1.94$ and $\varepsilon_{reo} = 1.76$ so that the length of the line can be calculated to be $L = \lambda_g/4 = 0.0368$ m where

$$\lambda_g = \frac{\lambda_{ge} + \lambda_{go}}{2} = 0.147174 \text{ m}.$$

The simulation results of the coupling response are shown in Figure 5.27(b) with Part 1 of the Software VNA connected to Port 1 of the coupler. The coupling strength at 1.5 GHz is -10.25 dB.

5.6.6 Rat-Race Hybrid Ring Coupler

Figure 5.28(a) shows the rat-race hybrid ring coupler (Pon, 1961), designed using the equations in Section 4.6.2 with a central frequency of $f_0 = 3$ GHz, $Z_0 = 50\,\Omega$ and $Z_a = \sqrt{2}Z_0 = 70.7\,\Omega$. The circuit consists of three transmission line sections of 25 mm in length and one section of 75 mm in length and four T-junctions. The coupler, Coupler_rrhr.nwk, is shown in Figure 5.28(b). The phase velocity of each transmission line section is taken to be $v_p = c$. With Port 1 of the Software VNA connected to Port 1 of the coupler and Port 2 of the Software VNA to Port 4 of the coupler, the coupling response can be observed. This coupling response is shown in Figure 5.28(c) with the frequency span from 1 to 5 GHz. The coupling strength at 3 GHz is -3 dB and the phase angle at 3 GHz is 90°. The 0.2 dB deviation bandwidth is 22.7% from 2.66 to 3.34 GHz. If Port 2 of the Software VNA is connected to Port 2 of the coupler, the coupling response to Port 2 of the coupler for an input from Port 1 can also be observed. The coupling strength at 3 GHz is again -3 dB and the phase angle at 3 GHz is $-90°$. The 0.2 dB deviation bandwidth is 18.7% from 2.72 to 3.28 GHz. If Port 2 of the Software VNA is connected to Port 3 of the coupler, it can be seen that Port 3 is very well isolated at 3 GHz.

Further examinations can be made with Port 1 of the Software VNA connected to Port 2, 3 or 4 of the coupler and Port 2 of the Software VNA to other ports of the coupler.

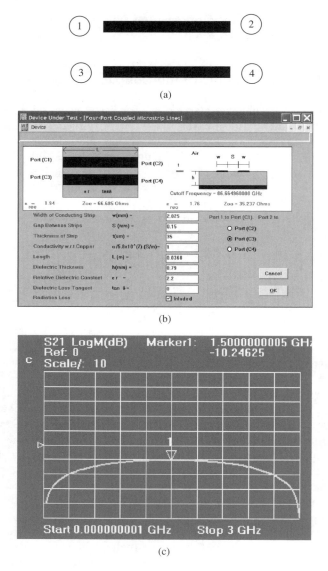

Figure 5.27   Microstrip coupled line coupler: (a) circuit diagram, (b) assembled circuit and (c) simulation results

5.6.7   March's Wideband Rat-Race Hybrid Ring Coupler

Figure 5.29(a) shows the wideband rat-race hybrid ring coupler (March, 1968), designed using the equations in Section 4.6.3 with a central frequency of $f_0 = 3\,\text{GHz}$ and $Z_0 = 50\,\Omega$. The circuit consists of three transmission

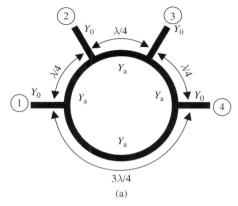

(a)

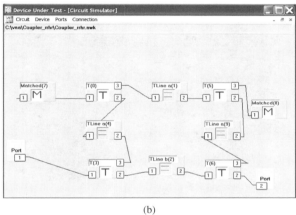

(b)

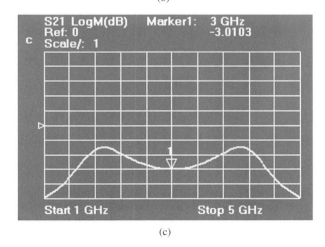

(c)

Figure 5.28 Rat-race hybrid ring coupler: (a) circuit diagram, (b) assembled circuit and (c) simulation results

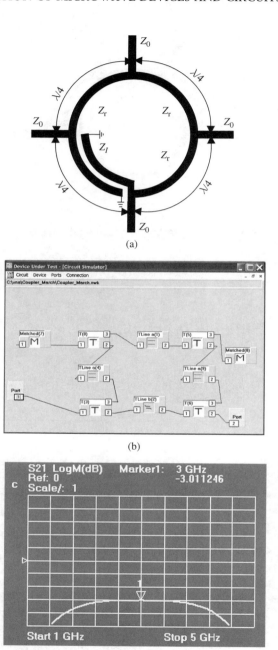

Figure 5.29 March's wideband rat-race hybrid ring coupler: (a) circuit diagram, (b) assembled circuit and (c) simulation results

line sections, a coupled line with two short-circuited terminations at two opposite ports and four T-junctions. The transmission line and coupled line sections have a length of 25 mm with the consideration of $v_p = c$. The characteristic impedance of the transmission line sections is $Z_r = 73\,\Omega$. The even and odd mode characteristic impedances of the coupled line are $Z_{0e} = (1 + \sqrt{2})Z_I = 176.2\,\Omega$ and $Z_{0o} = (\sqrt{2} - 1)Z_I = 30.2\,\Omega$, respectively. The coupler, Coupler_March.nwk, is shown in Figure 5.29(b). With Port 1 of the Software VNA connected to Port 1 of the coupler and Port 2 of the Software VNA to Port 4 of the coupler, the coupling response can be observed. It is shown in Figure 5.29(c) with the frequency span from 1 to 5 GHz. The coupling strength at 3 GHz is -3 dB and the phase angle at 3 GHz is 90°. The 0.2 dB deviation bandwidth is 56% from 2.16 to 3.84 GHz. The bandwidth is increased by 2.5 times compared with the rat-race hybrid ring design in Section 5.6.6. If Port 2 of the Software VNA is connected to Port 2 of the coupler, the coupling response to Port 2 of the coupler for an input from Port 1 can also be observed. The coupling strength at 3 GHz is again -3 dB and the phase angle at 3 GHz is $-90°$. The 0.2 dB deviation bandwidth is 62.6%. If Port 2 of the Software VNA is connected to Port 3 of the coupler, it can be seen that the isolation at 3 GHz is -69 dB.

Again further examinations can be made with Port 1 of the Software VNA connected to Port 2, 3 or 4 of the coupler and Port 2 of the Software VNA to other ports of the coupler.

5.7 FILTERS

5.7.1 Maximally Flat Discrete Element Low-Pass Filter

Figure 5.30(a) shows the design of a fifth order maximally flat low-pass filter with $Z_0 = 50\,\Omega$ and $f_c = 1$ GHz. For $N = 5$, the element values given in Section 4.8.3 and Table 4.4 for $\omega_c = 1$ are (Pozar, 1990; Matthaei, Young and Jones, 1965)

$g_0 = 1$, $g_1 = 0.618$, $g_2 = 1.618$, $g_3 = 2$, $g_4 = 1.618$, $g_5 = 0.618$, $g_6 = 1$.

Applying the impedance and frequency scaling tabulated in Table 4.7 gives

$$C_1 = g_1/(50 \times 2\pi f_c) = \frac{0.618}{100\pi \times 10^9} = 3.09 \times 10^{-12}\,\text{F},$$

$$L_2 = g_2 \times 50/(2\pi f_c) = \frac{1.618 \times 50}{2\pi \times 10^9} = 12.8 \times 10^{-9}\,\text{H},$$

$$C_3 = g_3/(50 \times 2\pi f_c) = \frac{2}{100\pi \times 10^9} = 6.366 \times 10^{-12}\,\text{F},$$

$$L_4 = L_2 = 12.8 \times 10^{-9}\,\text{H}$$

and

$$C_5 = C_1 = 3.09 \times 10^{-12}\,\text{F}.$$

Using these values, the assembled circuit, Filter_mfdelpf.nwk, is shown in Figure 5.30(b). With the selection of start frequency of 0 GHz and stop frequency of 2 GHz, the S_{21} response of the low-pass filter is shown in Figure 5.30(c). The response can be viewed with different scale values and frequency span. The cut-off frequency occurs at 1 GHz approximately.

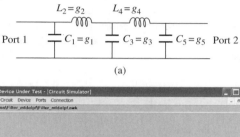

(a)

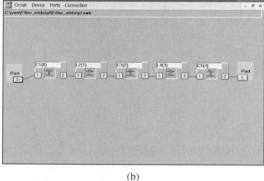

(b)

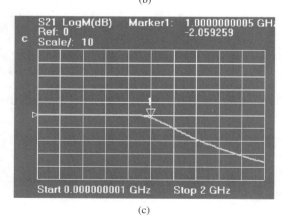

(c)

Figure 5.30 Maximally flat low-pass filter: (a) circuit diagram, (b) assembled circuit and (c) simulation results

5.7.2 Equal Ripple Discrete Element Low-Pass Filter

Figure 5.31(a) shows the design of a fifth order Chebyshev equal ripple low-pass filter with $Z_0 = 50\,\Omega$ and $f_c = 1\,\text{GHz}$ and ripple level of 0.5 dB. For $N = 5$, the element values given in Section 4.8.4 and Table 4.5 for $\omega_c = 1$ are (Pozar, 1990; Matthaei, Young and Jones, 1965)

$$g_0 = 1,\ g_1 = 1.7058,\ g_2 = 1.2296,\ g_3 = 2.5408,\ g_4 = 1.2296,$$

$$g_5 = 1.7058,\ g_6 = 1$$

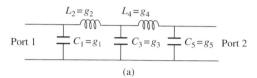

(a)

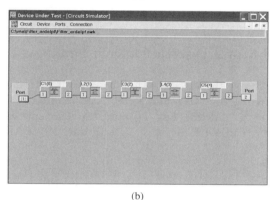

(b)

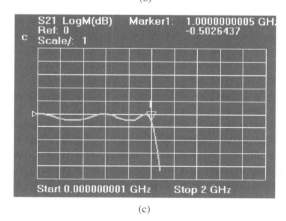

(c)

Figure 5.31 Equal ripple low-pass filter: (a) circuit diagram, (b) assembled circuit and (c) simulation results

so that the element values of the low-pass filter are

$$C_1 = g_1 / (50 \times 2\pi f_c) = \frac{1.7058}{100\pi \times 10^9} = 5.43\,\text{pF},$$

$$L_2 = g_2 \times 50 / (2\pi f_c) = \frac{1.2296 \times 50}{2\pi \times 10^9} = 9.785\,\text{nH},$$

$$C_3 = g_3 / (50 \times 2\pi f_c) = \frac{2.5408}{100\pi \times 10^9} = 8.09\,\text{pF},$$

$$L_4 = L_2 = 9.785\,\text{nH}, \quad C_5 = C_1 = 5.43\,\text{pF}.$$

Using these values, the assembled circuit, Filter_erdelpf.nwk, is shown in Figure 5.31(b). The simulated S_{21} response of the low-pass filter is shown in Figure 5.31(c) with the frequency span from 0 to 2 GHz. The cut-off frequency of the filter at $-0.5\,\text{dB}$ level occurs at 1 GHz. With the scale of 1 dB/Div, the ripple level of the filter can be observed.

5.7.3   Equal Ripple Discrete Element Bandpass Filter

Figure 5.32(a) shows the design of a fifth order Chebyshev equal ripple bandpass filter with $Z_0 = 50\,\Omega$, $f_0 = 3\,\text{GHz}$, bandwidth of 20% or 0.6 GHz and ripple level of 0.5 dB. For $N = 5$, the element values given in Section 4.8.4 and Table 4.5 for $\omega_c = 1$ are (Pozar, 1990; Matthaei, Young and Jones, 1965)

$$g_0 = 1, \ g_1 = 1.7058, \ g_2 = 1.2296, \ g_3 = 2.5408, \ g_4 = 1.2296,$$

$$g_5 = 1.7058, \ g_6 = 1.$$

Using the filter transformations in Tables 4.6 and 4.7,

$$L_{11} = L_{55} = 0.311\,\text{nH}, \quad C_{11} = C_{55} = 9.05\,\text{pF},$$

$$L_{22} = L_{44} = 16.31\,\text{nH}, \quad C_{22} = C_{44} = 0.1726\,\text{pF},$$

$$L_{33} = 0.209\,\text{nH and } C_{33} = 13.48\,\text{pF}$$

can be obtained. Using these values, the assembled circuit, Filter_erdebpf.nwk, is shown in Figure 5.32(b). The simulated S_{21} response of the bandpass filter is shown in Figure 5.32(c) with the frequency span of 2–4 GHz. Using the scale of 1 dB/Div, the equal ripple level bandpass response of the filter can be more clearly observed. The bandpass filter has a sharp cut-off on both sides of the pass band. The $-3\,\text{dB}$ bandwidth is 20% or 0.6 GHz approximately.

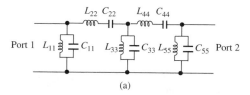

(a)

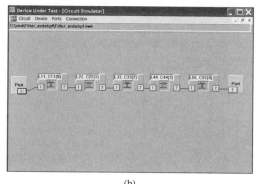

(b)

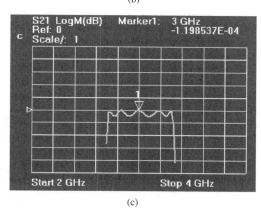

(c)

Figure 5.32 Bandpass filter: (a) circuit diagram, (b) assembled circuit and (c) simulation results

5.7.4   Step Impedance Low-Pass Filter

Figure 5.33(a) shows the design of an eighth order maximally flat low-pass filter with $Z_0 = 50\ \Omega$ and $f_c = 1\ \text{GHz}$. For $N = 8$, the element values given in Section 4.8.3 and Table 4.4 for $\omega_c = 1$ are (Pozar, 1990; Matthaei, Young and Jones, 1965)

$$g_1 = 0.3902, \quad g_2 = 1.1111, \quad g_3 = 1.6629, \quad g_4 = 1.9615,$$

$$g_5 = 1.9615, \quad g_6 = 1.6629, \quad g_7 = 1.1111$$

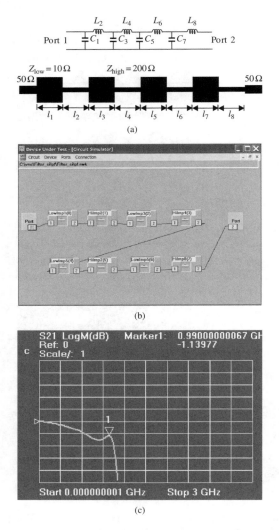

Figure 5.33 Microstrip step impedance low-pass filter: (a) circuit diagram, (b) assembled circuit and (c) simulation results

and

$$g_8 = 0.3902.$$

By choosing $Z_{high} = Z_h = 200 \, \Omega$ and $Z_{low} = Z_l = 10 \, \Omega$, the phase lengths of the high- and low-impedance sections can be calculated using the equations in Section 4.8.6, and they are tabulated in Table 5.1. All the phase lengths are less than $\pi/4$ or 90°. If the high- and low-impedance sections are realised using RT/Duroid 5880 with $\varepsilon_r = 2.2$, $h = 0.79$ mm and $t = 35\,\mu$m, the physical length of each section can be calculated using

$$l = \frac{(\beta l)\,\lambda_0}{2\pi\sqrt{\varepsilon_{re}}} = \frac{(\beta l)\,\dfrac{c}{f_c}}{2\pi\sqrt{\varepsilon_{re}}} = \frac{(\beta l) \times \dfrac{3 \times 10^8}{1 \times 10^9}}{2\pi\sqrt{\varepsilon_{re}}} = \frac{(\beta l) \times 0.3}{2\pi\sqrt{\varepsilon_{re}}}.$$

With the use of the Physical Transmission Lines: Microstrip Line device described in Section 3.15, the width of each section can be found based on its characteristic impedance requirement, and the ε_{re} value can be obtained accordingly. The physical length of each section can be chosen to meet the requirement. These values are tabulated in Table 5.1.

Using the parameters in Table 5.1, a step impedance low-pass filter, Filter_silpf.nwk, can be assembled as shown in Figure 5.33(b). The simulated S_{21} response of the bandpass filter is shown in Figure 5.33(c)

Table 5.1   Parameters of high- and low-impedance sections

No.	$Z_0(\Omega)$	βl (rad)	Air		RT/Duroid 5880 with $\varepsilon_r = 2.2$, $h = 0.79$ mm	
			l (mm)	ε_{re}	l (mm)	w (mm)
1	10	$\beta l_1 = \dfrac{C_1 Z_l}{Z_0} = \dfrac{g_1 Z_l}{Z_0} = \dfrac{0.3902 \times 10}{50}$ $= 0.07804 = 4.47°$	3.726	2.096	2.574	17.95
2	200	$\beta l_2 = \dfrac{L Z_0}{Z_h} = \dfrac{g_2 Z_0}{Z_h} = \dfrac{1.1111 \times 50}{200}$ $= 0.277775 = 15.92°$	13.263	1.610	10.453	0.04
3	10	$\beta l_3 = \dfrac{g_3 Z_l}{Z_0} = \dfrac{1.6629 \times 10}{50}$ $= 0.33258 = 19.06°$	15.88	2.096	10.968	17.95
4	200	$\beta l_4 = \dfrac{g_4 Z_0}{Z_h} = \dfrac{1.9615 \times 50}{200}$ $= 0.490375 = 28.10°$	23.414	1.610	18.453	0.04
5	10	$\beta l_5 = \dfrac{g_5 Z_l}{Z_0} = \dfrac{1.9615 \times 10}{50}$ $= 0.3923 = 22.48°$	18.731	2.096	12.938	17.95
6	200	$\beta l_6 = \dfrac{g_6 Z_0}{Z_h} = \dfrac{1.6629 \times 50}{200}$ $= 0.415725 = 23.82°$	19.849	1.610	15.644	0.04
7	10	$\beta l_7 = \dfrac{g_7 Z_l}{Z_0} = \dfrac{1.1111 \times 10}{50}$ $= 0.2222 = 12.73°$	10.61	2.096	7.329	17.95
8	200	$\beta l_8 = \dfrac{g_8 Z_0}{Z_h} = \dfrac{0.3902 \times 50}{200}$ $= 0.09755 = 5.59°$	4.658	1.610	3.671	0.04

with the frequency span of 0–3 GHz. Using the scale of 1 dB/Div, it can be seen that the response deviates slightly from the required maximally flat low-pass response, due to the approximations in the parameters.

5.7.5 Bandpass Filter Using Quarter-Wave Resonators

Figure 5.34(a) shows the design of a fifth order Chebyshev bandpass filter using quarter-wavelength short-circuited stubs with $Z_0 = 50\,\Omega$, $f_0 = 3\,\text{GHz}$, $\Delta = (f_2 - f_1)/f_0 = 20\%$ and ripple level of 0.5 dB. For $N = 5$, the element values given in Section 4.8.4 and Table 4.5 for $\omega_c = 1$ are (Pozar, 1990; Matthaei, Young and Jones, 1965)

$$g_0 = 1, \quad g_1 = 1.7058, \quad g_2 = 1.2296, \quad g_3 = 2.5408,$$

$$g_4 = 1.2296, \quad g_5 = 1.7058, \quad g_6 = 1.$$

At 3 GHz, $\lambda_0 = 0.1\,\text{m}$. If $v_p = c$ is chosen for all the transmission line sections so that $\lambda = \lambda_0$, the physical length of the transmission line sections is then $l = \lambda/4 = 0.025\,\text{m}$. Using the equation in Section 4.8.7, the characteristic impedance of each line section can be calculated as follows:

$$Z_{01} = Z_{05} = \frac{\pi Z_0 \Delta}{4 g_1} = \frac{\pi \times 50 \times 0.2}{4 \times 1.7058} = 4.6\,\Omega,$$

$$Z_{02} = Z_{04} = \frac{\pi Z_0 \Delta}{4 g_2} = 6.39\,\Omega \text{ and } Z_{03} = 3.09\,\Omega.$$

Using these values, the assembled circuit, Filter_bpfur.nwk, is shown in Figure 5.34(b). The simulated S_{21} response of the bandpass filter is shown in Figure 5.34(c) with the frequency span of 2–4 GHz. Using the scale of 1 dB/Div, the equal ripple level bandpass response of the filter can be more clearly observed. The ripple level is $-0.35\,\text{dB}$. The $-3\,\text{dB}$ bandwidth is 20% approximately. The bandwidth can be measured more accurately with a smaller frequency span or an increased number of points.

5.7.6 Bandpass Filter Using Quarter-Wave Connecting Lines and Short-Circuited Stubs

Figure 5.35(a) shows the design of an eighth order Chebyshev bandpass filter using quarter-wavelength connecting lines and short-circuited stubs with $Z_0 = 50\,\Omega$, $f_0 = 3\,\text{GHz}$, $f_1 = 1.95\,\text{GHz}$, $f_2 = 4.05\,\text{GHz}$, $d = 1$ and ripple level of 0.1 dB (Matthaei, Young and Jones, 1965). Using the procedure

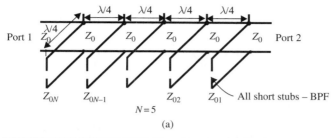

(a)

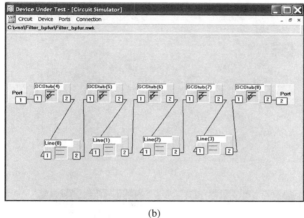

(b)

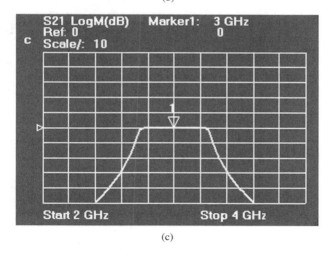

(c)

Figure 5.34 Bandpass filter using quarter-wave resonators: (a) circuit diagram, (b) assembled circuit and (c) simulation results

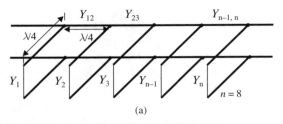

(a)

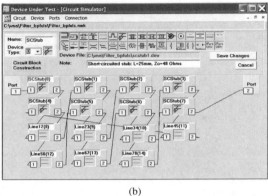

(b)

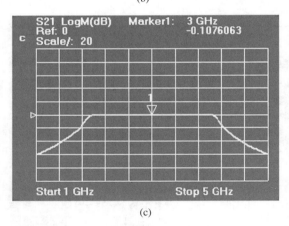

(c)

Figure 5.35   Bandpass filter using quarter-wave connecting lines and short-circuited stubs: (a) circuit diagram, (b) assembled circuit and (c) simulation results

described in Section 4.8.8, the characteristic impedances of the connecting transmission lines and stubs are

$$Z_{12} = Z_{78} = \frac{50}{1.288} = 38.8\,\Omega, \quad Z_{23} = Z_{67} = \frac{50}{1.364} = 36.7\,\Omega,$$

$$Z_{34} = Z_{56} = \frac{50}{1.292} = 38.7\,\Omega,$$

$$Z_{45} = \frac{50}{1.277} = 39.2\,\Omega, \quad Z_1 = Z_8 = \frac{50}{1.042} = 48\,\Omega,$$

$$Z_2 = Z_7 = \frac{50}{2.050} = 24.4\,\Omega,$$

$$Z_3 = Z_6 = \frac{50}{2.049} = 24.4\,\Omega \text{ and } Z_4 = Z_5 = \frac{50}{2.087} = 24\,\Omega.$$

At $f_0 = 3$ GHz, $\lambda_0 = 0.1$ m so that if $v_p = c$ is chosen for all the transmission line sections, the physical length of the transmission line sections is $l = \lambda/4 = 0.025$ m.

Using these values, the assembled circuit, Filter_bpfuls.nwk, is shown in Figure 5.35(b). The simulated S_{21} response of the bandpass filter is shown in Figure 5.35(c) with the frequency span of 1–5 GHz. Using the scale of 1 dB/Div, the equal ripple level bandpass response of the filter can be more clearly observed. The -3 dB bandwidth is 70% approximately.

5.7.7 Microstrip Coupled Line Filter

Figure 5.36(a) shows the design of a third order Chebyshev bandpass filter using microstrip coupled lines with $Z_0 = 50\,\Omega$, $f_0 = 3$ GHz, $\Delta = (f_2 - f_1)/f_0 = 20\%$ and ripple level of 0.5 dB. The filter has four coupled line sections. Using the procedure described in Section 4.8.9 and the Physical Coupled Lines: Two-Port Coupled Microstrip Lines device in Section 3.25, the parameters of the filter can be obtained as listed in Table 5.2 with the use of RT/Duroid 5880 with $\varepsilon_r = 2.2$, $h = 0.79$ mm and $t = 35\,\mu$m. The width and spacing of the microstrip coupled line are adjusted to meet the even and odd characteristic impedance requirements. The effective relative permittivity for the even mode and the odd mode is then obtained accordingly so that the physical length of each section can be calculated using the equation in Section 4.8.9.

Table 5.2 Parameters for microstrip coupled line filter made of RT/Duroid 5880

n	g_n	$Z_0 J_n$	$Z_{0e}(\Omega)$	$Z_{0o}(\Omega)$	w (mm)	S (mm)	ε_{ree}	ε_{reo}	Length (m)
1	1.5963	0.4436	82	37.7	1.508	0.093	1.91	1.72	0.018576
2	1.0967	0.2374	64.7	40.9	2.104	0.719	1.95	1.79	0.018294
3	1.5963	0.2374	64.7	40.9	2.104	0.719	1.95	1.79	0.018294
4	1	0.4436	82	37.7	1.508	0.093	1.91	1.72	0.018576

$w = 2.48$ mm for a line with $Z_0 = 50\,\Omega$

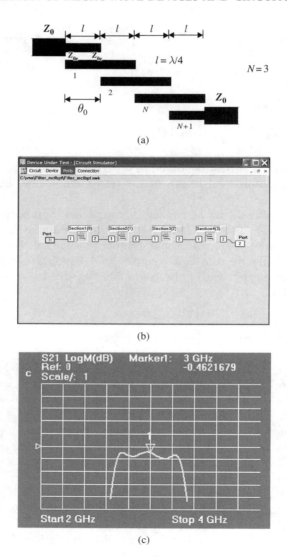

Figure 5.36 Microstrip coupled line filter: (a) circuit diagram, (b) assembled circuit and (c) simulation results

The assembled circuit, Filter_mclbpf.nwk, is shown in Figure 5.36(b). The simulated S_{21} response of the bandpass filter is shown in Figure 5.36(c) with the frequency span of 2–4 GHz. Using the scale of 1 dB/Div, the equal ripple level bandpass response of the filter can be more clearly observed, and the ripple level is 0.5 dB approximately. The filter has a loss of 0.5 dB, partly due to the ohmic loss on the conductors.

Table 5.3   Parameters for end-coupled microstrip line filter made of RT/Duroid 5880

n	g_n	Z_0J_n	$B_n(\times 10^{-3})$	C_n(pF)	θ_n(deg)	l_n (m)	Gap (mm)
1	1.5963	0.4436	11.046	0.586	142.7	0.028795	0.009
2	1.0967	0.2374	5.032	0.267	153.3	0.030934	0.025
3	1.5963	0.2374	5.032	0.267	142.7	0.028795	0.025
4	1	0.4436	11.046	0.586			0.009

5.7.8   End-Coupled Microstrip Resonator Filter

Figure 5.37(a) shows the design of the third order Chebyshev bandpass filter as that in Section 5.7.7, but realised using end-coupled microstrip lines. The required filter response parameters are $Z_0 = 50\,\Omega$, $f_0 = 3\,\text{GHz}$, $\Delta = (f_2 - f_1)/f_0 = 20\%$ and ripple level of 0.5 dB. Using the procedures described in Section 4.8.10, the Physical Transmission Lines: Microstrip Line device in Section 3.15 for the microstrip line and the Physical Line Discontinuities: Microstrip Line Discontinuities device in Section 3.20 for the gap, the parameters of the filter can be obtained as listed in Table 5.3 with the use of RT/Duroid 5880 with $\varepsilon_r = 2.2$, $h = 0.79\,\text{mm}$ and $t = 35\,\mu\text{m}$, in which

$$\theta_n = \beta l_n = \frac{2\pi}{\lambda_0}\sqrt{\varepsilon_{re}}l_n \text{ and } l_n = \frac{\theta_n\lambda_0}{2\pi\sqrt{\varepsilon_{re}}} = \frac{\theta_n \times 0.1}{2\pi\sqrt{\varepsilon_{re}}}.$$

For the microstrip line with $Z_0 = 50\,\Omega$, it can be obtained that w = 2.48 mm and $\varepsilon_{re} = 1.895$ at 3 GHz. The gap distance can be obtained by using the S_{11} display on the 2PChart to match the required capacitance value, in which the admittance value is $Y_n = jB_n/100$.

The assembled circuit, Filter_ecmrbpf.nwk, is shown in Figure 5.37(b). The simulated S_{21} response of the bandpass filter is shown in Figure 5.37(c) with the frequency span of 2–4 GHz. Using the scale of 1 dB/Div, the equal ripple level bandpass response of the filter can be more clearly observed, and the ripple level is again 0.5 dB approximately. The filter has a loss of 0.5 dB, partly due to the ohmic loss on the conductors.

5.8   AMPLIFIER DESIGN

5.8.1   Maximum Gain Amplifier

Figure 5.38(a) shows the design of an amplifier with maximum gain at 2 GHz using the principle described in Section 4.9.1. The transistor used is a Siemens CGY50 FET (Siemens, CGY50 FET datasheets), and the

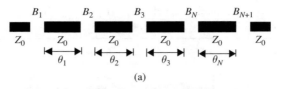

(a)

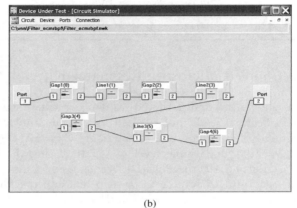

(b)

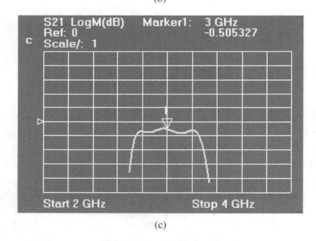

(c)

Figure 5.37   End-coupled microstrip filter: (a) circuit diagram, (b) assembled circuit and (c) simulation results

S-parameters of the FET at different frequencies with respect to $Z_{0,\text{ref}} = 50\,\Omega$ are listed in Table 5.4.

At 2 GHz, it can be calculated that

$$|\Delta| = 0.332 < 1 \text{ and } K = 1.742 > 1.$$

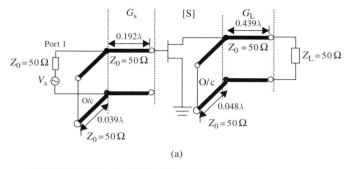

(a)

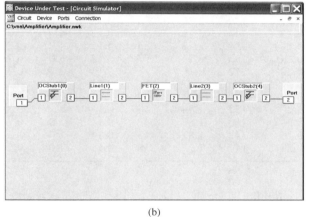

(b)

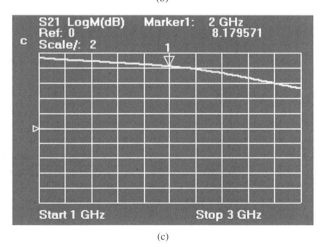

(c)

Figure 5.38 Maximum gain amplifier: (a) circuit diagram, (b) assembled circuit and (c) simulation results

Table 5.4 S-parameters of Siemens CGY50 FET ($V_D = 4.5$ V, $V_G = 0$, $Z_0 = 50$ Ω

	S_{11}		S_{21}		S_{12}		S_{22}	
f (GHz)	Mag	Ang	Mag	Ang	Mag	Ang	Mag	Ang
0.2	0.25	−31	3.30	164	0.14	5.0	0.05	−144
0.4	0.27	−34	3.20	158	0.14	0.0	0.05	−133
0.6	0.21	−44	3.17	150	0.13	−2.0	0.08	105
0.8	0.20	−54	3.09	142	0.13	−3.0	0.01	91
1.0	0.19	−65	3.00	134	0.13	−4.0	0.12	81
1.2	0.18	−77	2.90	126	0.13	−5.0	0.14	74
1.4	0.18	−93	2.81	118	0.13	−5.0	0.16	68
1.6	0.17	−103	2.70	111	0.13	−6.0	0.17	62
1.8	0.17	−119	2.60	103	0.13	−5.0	0.18	56
2.0	0.17	−130	2.50	96	0.12	−5.0	0.19	51
2.2	0.18	−141	2.42	94	0.12	−4.0	0.20	46
2.4	0.18	−152	2.33	83	0.12	−4.0	0.21	42
2.6	0.19	−163	2.24	77	0.12	−3.0	0.21	39
2.8	0.20	−172	2.16	71	0.13	−3.0	0.21	36
3.0	0.21	179	2.07	65	0.13	−2.0	0.21	33
3.2	0.22	172	2.01	60	0.13	−2.0	0.21	30
3.4	0.23	162	1.94	54	0.13	−2.0	0.21	29
3.6	0.24	153	1.87	49	0.14	−1.0	0.21	28
3.8	0.26	148	1.81	43	0.14	−1.0	0.21	27
4.0	0.28	142	1.75	38	0.15	−1.0	0.20	27

Hence the amplifier is stable. The required reflection coefficients Γ_s and Γ_L to achieve maximum gain are, respectively,

$$\Gamma_s = 0.124\angle 124.8° \text{ and } \Gamma_L = 0.154\angle -54.8°.$$

The corresponding input impedances at the source and load ends are

$$Z_s = 42.532 + j8.836 \,\Omega \text{ and } Z_L = 57.678 - j14.819 \,\Omega.$$

Using the impedance transformation equations in Section 4.9.1, the length of the open-circuited stub and the length of the transmission line at the source end are 0.039 λ and 0.192 λ, respectively, where λ is the wavelength at 2 GHz, or 5.85 and 28.8 mm, respectively, if $v_p = c$. Similarly the length of the open-circuited stub and the length of the transmission line at the load end are 0.048λ and 0.439λ, respectively, or 7.2 and 65.85 mm, respectively, if $v_p = c$ The maximum transducer gain of the amplifier is

$$G_{T\max} = 1.016 \times 6.25 \times 1.036 = 6.576 \text{ or } 8.18 \text{ dB}.$$

Using the above parameters, the assembled amplifier circuit, Amplifier_maxg.nwk, is shown in Figure 5.38(b). The S-parameters of the

FET are input using User-Defined S-Parameters: Two-Port Device described in Section 3.35. The simulated S_{21} response of amplifier is shown in Figure 5.38(c) with the frequency span of 1–3 GHz. The amplifier is very well matched at 2 GHz at both input and output ends, which can be seen using S_{11} and S_{22} displays. The gain at 2 GHz can be located using Marker 1 to be 8.18 dB, as calculated above. This gain is, however, not the largest in the band of display as the result of the nonlinear S_{21} response of the FET.

For the purpose of simulating a balanced amplifier, use the SaveData function of the Software VNA to save the S-parameters of the maximum gain amplifier over the frequency range from 1 to 3 GHz as amp1.dat.

5.8.2 Balanced Amplifier

Figure 5.39(a) shows the design of a balanced amplifier using the same FETs (Siemens, CGY50 FET datasheets) as that in the maximum gain amplifier of Section 5.8.1 and the configuration of Figure 5.39(b). The balanced amplifier also operates at the central frequency of 2 GHz. The 3 dB Wilkinson power dividers, as described in Section 4.1.1, are made of air lines with $v_p = c$ so that the length of the quarter-wave line section is 37.5 mm. The quarter-wave transformer lines are also made of air lines with characteristic impedance of 50 Ω. The amplifiers, Amp1 and Amp2, are made of the maximum gain amplifier of Section 5.8.1 with the S-parameters saved in amp1.dat.

The assembled balanced amplifier circuit, Amplifier_balan.nwk, is shown in Figure 5.39(b). The S-parameters of the maximum gain amplifier of Section 5.8.1 in amp1.dat are imported using the User-Defined S-Parameters: Two-Port Device described in Section 3.35. The 3 dB Wilkinson power dividers are defined in the same way as Section 5.5.1, except that the quarter-wave line length is 37.5 mm.

The simulated S_{21} response of the amplifier is shown in Figure 5.39(c) with the frequency span of 1–3 GHz. The amplifier is very well matched at 2 GHz at both input and output ends, which can be seen using S_{11} and S_{22} displays. The gain at 2 GHz is the same as the maximum gain amplifier of Section 5.8.1, i.e. 8.18 dB. Compared with the S_{21} response shown in Figure 5.38(c) for the maximum gain amplifier, the balanced amplifier has a smoother response and wider bandwidth.

5.9 WIRELESS TRANSMISSION SYSTEMS

5.9.1 Transmission Between Two Dipoles with Matching Circuits

Figure 5.40(a) shows a wireless transmission system with two dipole antennas and single-stub matching circuits for both antennas operating

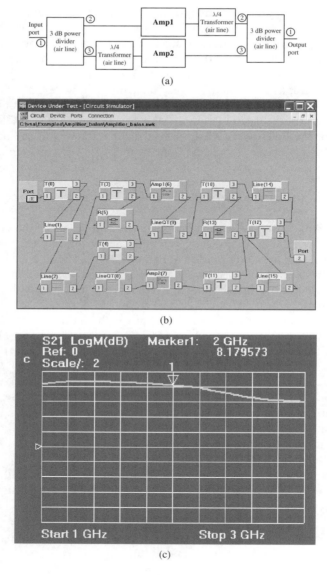

Figure 5.39 Balanced amplifier: (a) circuit diagram, (b) assembled circuit and (c) simulation results

at 3 GHz. The system can be simulated using transmission lines, short-circuited stubs and the Antennas: Transmission between Dipole Antennas device described in Section 3.32. The assembled circuit using the devices, System_2dip.nwk, is shown in Figure 5.40(b). The length of the dipoles is chosen to be 50 mm and the wire radius to be 0.5 mm. The matching circuits

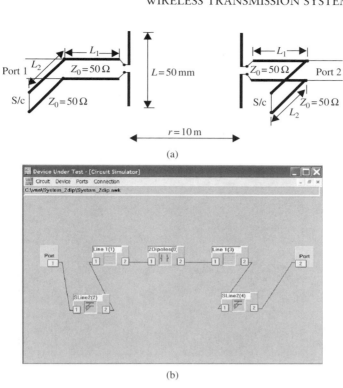

(a)

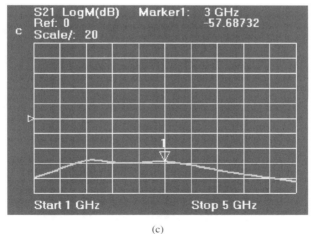

(b)

(c)

Figure 5.40 Transmission between two dipoles: (a) circuit diagram, (b) assembled circuit and (c) simulation results

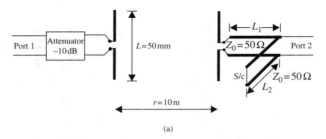

(a)

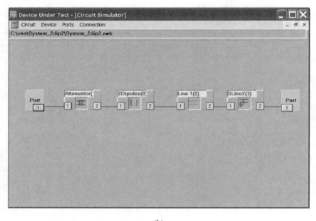

(b)

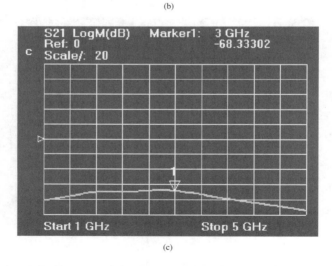

(c)

Figure 5.41 Transmission between two dipole antennas with an attenuator: (a) circuit diagram, (b) assembled circuit and (c) simulation results

have been described in Section 5.2.2. The result of the S_{21} response for a separation distance of 10 m is shown in Figure 5.40(c) with the frequency span of 1–5 GHz. The transmission loss at 3 GHz is 57.7 dB. Both dipoles are very well matched at 3 GHz, which can be observed using S_{11} and S_{22} displays.

5.9.2   Transmission Between Two Dipoles with an Attenuator

Figure 5.41(a) shows a wireless transmission system with two dipole antennas, an attenuator in the transmitting end and a single-stub matching circuit in the receiving end. The system is designed to operate at 3 GHz. This circuit is a modification of that shown in Figure 5.40(a) in which the matching circuit in the transmitting end is replaced by an attenuator. The assembled circuit using the devices, System_2dip2.nwk, is shown in Figure 5.41(b). The length of the dipoles is again chosen to be 50 mm and the wire radius to be 0.5 mm. The matching circuit is the same as that in Section 5.2.2. The result of the S_{21} response for a 10 m separation distance and 10 dB attenuation in the attenuator is shown in Figure 5.41(c) with the frequency span of 1–5 GHz. The transmission loss at 3 GHz is 68.3 dB. The dipole in the transmitting end is well matched over a wide band due to the use of a 10 dB attenuator, and the dipole in the receiving end is very well matched at 3 GHz. These can be observed using S_{11} and S_{22} displays.

REFERENCES

Cohn, S.B. (1968) 'A class of broadband three-port TEM-mode hybrids', *IEEE Transactions on Microwave Theory and Techniques*, **MTT-16**, 110–6.

March, S. (1968) 'A wideband strip line hybrid ring', *IEEE Transactions on Microwave Theory and Techniques*, **MTT-16**, 361.

Matthaei, G.L., Young, L. and Jones, E.M.T. (1965) *Microwave Filters, Impedance Matching Networks and Coupling Structures,* McGraw-Hill, New York.

Pon, C.Y. (1961) 'Hybrid ring directional coupler for arbitrary power division', *IRE Transactions on Microwave Theory and Techniques*, **MTT-9**, 529–35.

Pozar, D.M. (1990) *Microwave Engineering*, Addison-Wesley, New York.

Index